Greg,
More than
you need about
my journey.
I enjoy our discussions,
Pete

Steerageway

HOW EMPLOYEE OWNERS SET DIRECTION
AMID THE CURRENTS OF CHANGE

PETER S. STRANGE

keen
custom
media

Introduction

Those who have sailed, canoed, powerboated, or rafted are all informed about one of nature's realities: If you drift with the current, you will go wherever the current takes you. No matter how good you are as a person, no matter how clear your vision and goals; you have no choice in the matter.

The only way that you can impose your will upon nature is to create steerageway. You must use the wind, or your paddle, or your engine to create a speed different from the current around you. Only then can you set direction toward your desired destination.

It is the same in business and in life. If you drift with the currents, you will have no choice in your path or your destination.

This is a book about people who used their ideas, their energy, and their love to create steerageway and to set direction for a company amid the currents of change. The story is about Messer as a company, but make no mistake—the subject of the book is leadership. (Messer—the name we employee owners use when talking about our company—will be used throughout this book. I will not attempt to differentiate between Messer Construction, Messer, Inc., or our many predecessors and affiliates.)

The story will follow a timeline but only because history gives context to the choices and actions of the people. It was the people who made the difference. They created their own story, rather than allowing the times, the pressures, or the opportunities to dictate it for them. Each of these people, in his or her own way, created steerageway and set a direction that impacted others far into the future.

One goal for this book is to share some strategies that may help Messer leaders of the future create steerageway for the company and themselves. More important to me is the opportunity to share the stories of some amazing people who met the test of leadership—they helped others get better results, make better decisions, and think longer term.

Be advised that this is not a history book. It will not stand the cold, hard test of verifiable data. I share the tribal stories that have been handed down to me. As happens with tribal stories, they were changed by the tellers—even when they had access to the facts—and they have been changed by me as I have sought to capture, embrace, and share the heart of the matter. I have told no purposeful lies; I simply care less about the facts than I care about the lessons to be learned. You are welcome to your own version of what happened; I stand by my version of what mattered.

I add a second disclaimer. I have not shared stories of bad people. I have met only a few bad people, and they contributed nothing to my understanding so I have chosen to omit them from the book. Where I have included bad choices or bad results, it is not because the people were bad, it is to add to our learning and to acknowledge that #*#* does indeed happen. To write about leadership without acknowledging failure would call into question the entire effort.

Finally, I make no claim to original thought. Therefore, I elect to use the collective "we" rather than the pompous "I" when sharing ideas not directly attributable to others. I do so not from false humility but out of fear that the value of the ideas themselves may be eroded in the fight for ownership. It is enough to share the stories; it would be too much for any of the participants to claim sole ownership.

Action, not adulation, is what matters most. At the end, I will tell stories about closing the gaps. What you do to create steerageway will be the test of the sharing.

Keith's Theater Building

Big projects for a start-up company:
Cox and Shubert Theaters (left) and Dixie
Terminal (above).

moved their offices to Burnet Avenue near Reading Road (current site of Park Tower), and the story of Messer officially began.

A PERSONAL ASIDE

My grandpa and his brother worked for the Ohio Building Company as carpenters for almost two years. They worked on Dixie Terminal, the Cox & Schubert Theater Buildings, and the Keith's Building. In the late 1960s when the Keith's Building was being torn down to make room for urban renewal, Grandpa insisted that I go with him one night to watch. He wanted me to see how hard it was to kill a well-built building, and how sad it was when one died. Later, when the first building I worked on for Messer—Sander Hall on the University of Cincinnati campus—was imploded, I rented a parking lot and brought my children to see the "event." For me, it was as sad as a public hanging.

Messer began with ten thousand dollars in capital, two strong leaders, and a lot of entrepreneurial will. The company was challenged to find good work as the economy struggled; nonetheless, they grew their staff. Four events defined those early years. The first was the addition in 1932 of the people from Wells Brothers, a large industrial contractor in Chicago that failed during the Depression. The Wells Brothers' vice president of operations, Earl Wheeler, took a job with Messer, and he brought with him a number of very experienced builders. Those builders provided the capacity for Messer's early growth.

The second event was the successful bid on the Nashville Post Office Project (now the Frist Center for the Arts). By late 1932, the company was struggling to maintain cash flow. They had good resources, but there just was not enough commercial construction to engage their capacity. The company identified the one project they had to get to survive—the Nashville Post Office. Mr. Wheeler put the bid together, added up the final recap, and then deducted 2 percent from total cost. Messer was the successful bidder; and they held their first office party when the project budget reached breakeven status.

THE FOUNDATION OF ETHICS

The bid for the Nashville Post Office Project was sent by telegraph to the Post Office Department in Washington, D.C. After the bids were opened and Messer was low bidder, Frank Messer, accompanied by a very young lawyer named Leonard Weakley, took the train to Washington to sign the contract. Mr. Weakley (all of these folks were so old when I came along that I can't even

Miss Helen Shrenk (center): a very small person
with a very large influence.

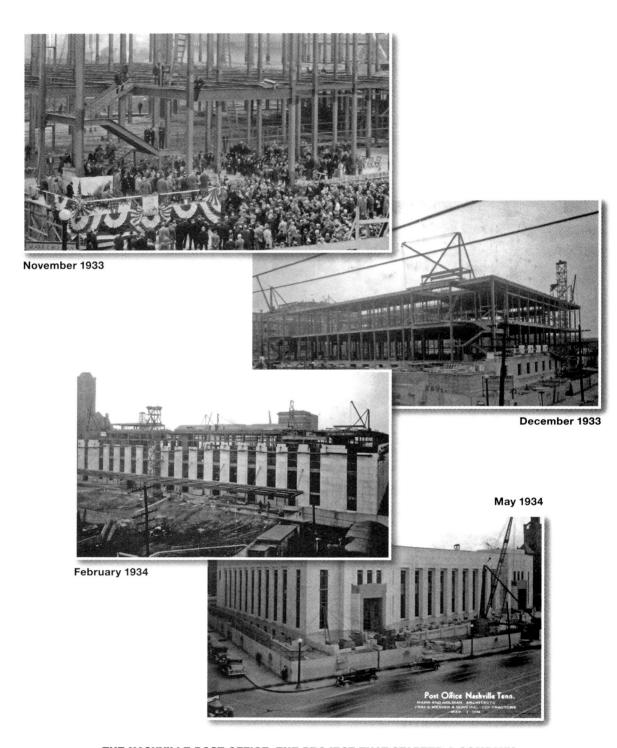

November 1933

December 1933

February 1934

May 1934

THE NASHVILLE POST OFFICE: THE PROJECT THAT STARTED A COMPANY.

Wright Aeronautical under construction

2

THE GROWTH YEARS
[1937–1961]

The Players
EARL WHEELER
HAROLD THATCHER
D. A. (POP) JOHNSON
JIM MATHEWS

Earl Wheeler graduated from the University of Illinois with a degree in civil engineering, but first and foremost he was an accomplished businessman, able to forge relationships with other business leaders of his day. He chaired the Cincinnati Chamber of Commerce Board, he was active in business clubs, and he had personal friendships with the CEOs of major companies. Under his leadership, Messer would become one of the largest building contractors in the United States.

Mr. Wheeler took over leadership of Messer when America was coming out of the Depression and gearing up for war. Although the U.S. didn't actually enter the war until 1941, by 1939 the economy was growing and production was booming in nearly every industrial category. Messer had the capacity and the leadership to serve the boom. Through the forties and fifties, Messer traveled across the eastern half of the United States, serving large, repeat customers.

Among the national customers that Messer served were Federated Department Stores, Neisner's stores, Schenley's in Indiana and Pennsylvania, National Distillers in Cincinnati, many of the independent distillers throughout

Kentucky—and, of course, the federal government. With his contacts and his abilities, Mr. Wheeler was successful in negotiating contracts without bidding, resulting in projects as far south as Texas and Alabama, and as far east as Boston.

And Messer was a leading government contractor, completing dozens of projects at facilities like Wright-Patterson Air Force Base in Dayton, Ohio, Avon and Bluegrass Army Depots in central Kentucky, and many others. A measure of Messer's size and capacity is the fact that when Wright Aeronautical (now General Electric's Evendale, Ohio, facility) was awarded to Messer by the U.S. government in 1940, it was, at approximately $40 million, the largest contract ever awarded to a single contractor in the history of the country. In contrast, the Boulder Dam project, awarded just a few years before, required a consortium of six contractors—including Bechtel, Kaiser, Morrison-Knudsen, and Kiewit—at approximately $49 million.

One reason Messer was able to compete for large, complex projects was because they had a unique resource in estimating and planning. That resource was Mr. Horace Thatcher. Mr. Thatcher was one of those people for whom no detail was too small. It was said that project managers and superintendents regularly ordered materials directly from his estimating recaps without ever consulting the plans or specs—he was just that good. Even the subcontractors relied upon "Thatch" to validate their own quantity surveys and to check their shop drawings. This ability to create a reliable plan and execute from that plan set Messer apart from its competition. Mr. Thatcher fit perfectly with Mr. Wheeler; neither of them could conceive of wasting even one nail, which was a good attitude to have coming out of the

Wright Aeronautical Project rendering: Thirty-three acres under roof; in 1940 the largest project let to a single contractor by the U.S. government in history.

Depression and entering into a period of profound shortage as a result of the war. Part of the ability to negotiate work during this period was based upon convincing customers that Messer was the company that could do more with less.

Mr. Thatcher kept a collection of pencil stubs neatly standing in a row on the head of his office door casing. He kept them there to remind others that you could use a pencil down to one inch and that wasting pencil lead was surely a symptom of prevailing waste.

COMMUNICATION IS NOT WHAT YOU SEND, BUT WHAT IS RECEIVED

Later in his career, when Charles Messer was pricing work in Estimating, Mr. Thatcher tried an experiment in helping his friends in the field. While Thatch could not stand waste, he also believed that every job plan should recognize the reality of the work. Mr. Thatcher believed that Charles Messer did not respect the differences between individual projects; that for Charles brickwork at ground level was the same as brickwork in the penthouse, ignoring the challenges of logistics and location.

So when Messer was proposing on a project that was fourteen floors above ground level, Mr. Thatcher decided that he would balance out Charles's lack of perception. Thatch added into the estimate two thousand common bricks, even though there was no brickwork in the project; his theory being that the money Charles

Wright Aeronautical in 1942

Wright Aeronautical producing engines for the war effort.

put in for the bricks would offset the money he left out for logistics and location on the rest of the materials. Thatch's plan worked. Messer got the job, even though there was a little extra money in the estimate; Thatch quietly celebrated his ability to fill the gaps so that the builders in the field had the right amount of resources for the project.

Unfortunately, Mr. Thatcher did not communicate his plan to the people in the field, not thinking about the possibility that they might order materials from the estimate. In due course, Thatch received a call from the project manager who said, "Thatch, we've carried two thousand bricks up fourteen flights of steps; now what are we supposed to do with them?"

Communication is not your plan, your creativity, or your delivery; communication is whatever is acted upon by those on the receiving end.

When the U.S. entered World War II, capacity became the name of the game, and Messer had capacity. The generation of project leaders who came from Wells Brothers during the Depression had a lot of experience, and most of them were too old for the draft. Leaders like Pop Johnson, Walter Wallner, Bill Raidt, Ed Willing, and Jim Mathews were capable of leading projects of any size, and Mr. Wheeler was the man who could get the projects. Work during the war was primarily for the government and was 100 percent union. The problem was that there was a severe shortage of labor. To cover the gap, many of the Messer leaders brought their sons, nephews, brothers, and cousins to the job—leading to the multigenerational environment we have today. Walter's son, Judd, went on from Messer to be the project manager on the Assembly building at Cape Canaveral (then the largest enclosed structure in history in terms of cubic feet). Bill Raidt's son, Bill Jr., went on to be Transportation Secretary of the State of Ohio. And, Dick Glassmeyer's younger brother, Joe, who joined the company as a sixteen-year-old water boy, went on to be the heart and soul of Messer.

LABOR RELATIONS: A DIFFERENT WORLD
When skilled labor was scarce and unions were strong, part of the job of a project leader was dealing with union issues. Most issues arose over jurisdictional disputes: a claim by one union that another union was doing their work. Jurisdiction was a matter of

tradition, negotiation, and bullying, and could add expense and complications to fairly simple tasks. For instance, for many years, both the carpenters' and the plumbers' unions claimed jurisdiction over the simple task of putting in blocking to support toilet-room accessories. The amount of time spent arguing over this task often exceeded the time it would take to install the blocking.

Jim Perin shared the story of one jurisdiction dispute. Jim Mathews was the project manager for one of the large buildings at Wright Aeronautical. When it came time to remove the material hoist from the building—jurisdiction claimed by the ironworkers' union—Jim Mathews had some laborers come in over the weekend and take down the hoist. On Monday, the entire leadership of the ironworkers' union showed up and announced that they were calling a general strike to shut down the project.

Jim Mathews met them at the gate and asked what reason they might have for calling a strike. They said that he knew the reason since he had caused the problem by assigning their jurisdiction to common laborers. (This was also a time when race relations were so bad that the Wright Aeronautical unions struck over having to work with African Americans; and the laborers were mostly African American.) Jim Perin, as a young project engineer on the project, believed that all hell was about to break loose so he stayed to see the fight. But, instead of yelling or explaining, Jim Mathews responded, "Gentlemen, you are wrong. No one took your jurisdiction because no one took down the material hoist."

When the ironworkers' reps tried to shout him down, Jim Mathews simply responded, "Come with me and I will prove the hoist is still there."

Jim Perin followed along because he had personally supervised the loading out of the hoist on Sunday. He could not imagine what would happen when the union delegation saw that the hoist was indeed gone.

When they rounded the corner of the building, before anyone else could say anything, Jim Mathews shouted, "Oh my God, somebody stole our hoist! Jim, call the police. We need to get

Foley's Department Store, Houston, Texas. Formwork and rebar for the longest span (250 feet) concrete beam in history (left).

FOLEY BROS., HOUSTON, TEXAS
KENNETH FRANZHEIM, ARCHITECT
EDWARD E. ASHLEY, CONSULTING ENGINEER
RAYMOND LOEWY ASSOCIATES, STORE DESIGNERS
FRANK MESSER & SONS, INC, CONTRACTORS
MARCH 15, 1947   PHOTO....57

Springfield, Massachusetts

Buffalo, New York

Cincinnati, Ohio

Boston, Massachusetts

NEISNER'S VARIETY STORES. MESSER BUILT ACROSS THE EASTERN HALF OF THE COUNTRY.

them to investigate!" At which point the ironworkers' union reps shook their heads and left the site.

Not all union disputes were settled so easily, and there was plenty of gamesmanship on all sides, but this was one time when the project manager won. Project leaders, then and now, must be good planners, good communicators, good leaders, and occasionally good performers.

After the war, construction boomed as American industry moved from war production to consumer goods. Housing also grew to accommodate the veterans returning to civilian life. In 1947, Messer began construction of the Terrace Hilton Hotel building, one of the largest buildings in downtown Cincinnati. Dick Glassmeyer was the "timekeeper"—the site-based book-keeper for the project. Dick would later become secretary of the company. In 1954, Mr. Wheeler negotiated the Procter & Gamble General Office Building Project in downtown Cincinnati. Pop Johnson was the project manager and Jim Perin was assistant project manager. That project was the start of a more-than-twenty-year partnership between Messer and P&G.

AN OPPORTUNITY LOST

Henry Parker of Stanford University published a book in 1974 titled *Construction Productivity*. In his introduction, Dr. Parker stated that the seminal work on construction productivity had taken place on the Procter & Gamble General Office Project in 1954. When I read those words one evening at home, I literally could not wait to get to the office the next day. Messer was the general contractor on that P&G project, and Jim Perin had been the project engineer! I was sitting in Jim Perin's office when he arrived the following morning. "Jim," I shouted, "how could I not know about this? Messer was the leader!"

After Jim got me calmed down, he told me this story. P&G had in-vited the entire project team to a six-week planning session in Flor-ida. The team broke the project down into discrete elements and set a productivity goal for each element. They created scorecards and tracked the elements through to completion. They called their approach the Florida System. Jim was convinced that the process saved significant dollars and cut months off the schedule.

"So what happened?" I asked.

"Well," said Jim, "I made an appointment to see Mr. Wheeler (no one saw Mr. Wheeler except by appointment) when the project was complete. I explained the process to him, and he told me that he appreciated my enthusiasm but he would set the priorities for Messer. And, that was the end of the 'Florida System.'"

This is one of the saddest business stories I have ever heard. Messer was in the lead but allowed self-satisfaction (or something) to get in the way of needed change. And Jim, who valued having a job above all else, didn't even consider quitting over this stupidity. By the 1980s, P&G had found contractors who were willing to change, and Messer lost the account. It took more than twenty years for Messer to re-earn P&G's confidence. If today we seem obsessed with change, it is because we never again want to voluntarily give up the lead.

In spite of an aversion to change, Messer found plenty of opportunities to grow capacity as veterans returned home. In general, these workers were like Joe Glassmeyer. They had entered the workforce or the military very young; they were smart, energetic, and wanted new opportunities. One of those young men was Alfred C. Berndsen. Al returned from the service and entered the University of Cincinnati on the GI Bill. When it came time to co-op (the career-related work experience required then as now by U.C. as part of its co-operative education model), Al informed the co-op coordinator that he had a wife and a child and could not afford a low-paying co-op job, so he would make his own arrangements for work. Al's father was a small contractor/developer on the west side of Cincinnati, and Al had worked for his father as a carpenter before going to the service. Al talked with some of his father's friends, who got him a co-op job as a carpenter with Messer working in the crew. Al had no intention of working for Messer after graduation, but craftwork paid the bills and he built friendships with his fellow workers. As graduation neared, Al interviewed with several companies, ultimately accepting a job with Chrysler Corporation. He was about three weeks away from moving his family to a new city when he decided that he should visit some of his old friends at a new project Messer was starting downtown. The project was the Cincinnati Gas & Electric Company Annex Building. Al went down just to look in the hole and say goodbye to friends.

When Axel Gustaffsen, the carpenter foremen, saw him he roared, "Kid, where have you been? We need help with layout!"

The Terrace Plaza Hotel: A major post-war project and a Cincinnati icon containing artwork by Calder, Steinberg, and Miro.

The Terrace Plaza, a fully riveted structure, September 1946 (above) and March 1947 below. (A worker was killed by a hot rivet dropped from many floors above him.) The completed building in July 1948 (right).

Al explained that he wouldn't be coming back because he had taken a job with Chrysler. Axel and the laborer foreman, Alt Turner, took Al in to see Pop Johnson and told Pop that he had better give Al a job because they needed the help. Pop offered Al a job because his friends supported him; Al accepted the job out of friendship.

Al then had to go home and explain to his wife that they weren't moving (which she liked very much) but that the new job didn't pay as well as Chrysler (which she did not like at all). She sent Al back to ask for a raise after only two weeks on the job. Probably out of fear of Al's wife, Pop gave him the raise. The day the raise went into effect, Al was called into Mr. Wheeler's office. Mr. Wheeler looked down from his chair on the dais and informed Al that Pop Johnson had no authority to hire him, and, furthermore, Pop sure as hell didn't have the authority to grant him a pay raise. Al was already thinking about what he would say to his wife to explain this turn of events, when Mr. Wheeler stood up and said, "You had better be worth it." With that, Messer hired its future leader.

WHY DID THEY STAY?

Joe Glassmeyer was working as a project manager in the late 1950s, overseeing several renovation projects in downtown Cincinnati. Because he needed to move from job to job, Joe quickly ran up almost thirteen dollars in parking meter charges. Being new to management—and being on a strict budget because he had four kids at the time—Joe sought expense reimbursement from the company, which resulted in his being called into Mr. Wheeler's office. Mr. Wheeler informed Joe that being a project manager was a very responsible job, which meant that he would be responsible for his own expenses—if he expected to keep the job. Joe elected to keep the job. Mr. Wheeler deposited Joe's expense report in the trashcan before Joe left his office. Messer employees didn't stay for the pay, and they didn't stay because they were treated kindly. They must have stayed for each other. They became a team with shared purpose and a sense of accomplishment. The sense of accomplishment came from building great buildings; the shared purpose may have been "to survive the boss."

In spite of all of his talents, Mr. Wheeler did not respect his own mortality. He was a man alone at Messer. His office door was closed, his desk was on a dais, like a raised platform for a throne, and he certainly didn't want or need anyone else from Messer in his circle of business contacts.

June 1952

April 1951

July 1952

June 1953

CINCINNATI GAS & ELECTRIC ANNEX: THE PROJECT WHERE CRAFT WORKERS HIRED AL BERNDSEN.

He did not share information, exposure, nor, if the stories are true, money. In 1962, while on a fishing trip with some friends (no one from Messer), Mr. Wheeler developed a severe medical condition. He died shortly after his return . . . and all of his influence, his contacts, and his information died with him. No one else at Messer had the customer relationships nor the information needed to step in and lead the sales process. If today we seem to obsess about having two people in every meeting, and about sharing everything we know with everyone we touch, it is because we want to avoid losing the ground we worked so hard to gain. Mr. Wheeler took Messer to national stature, but when he died, the company was not able to stay in that place. It took almost forty years to regain the position in the industry that we once had.

Mr. Wheeler grew Messer from 1937 through 1961. He created project opportunities and relationships that died when he died; but he (and circumstances) brought together a group of builders who would continue the story for decades to come.

Garfield Tower

3

CHANGE OF DIRECTION
[1962–1975]

The Players
CHARLES M. MESSER
LESLIE L. SUNDAHL
ALFRED C. BERNDSEN
LOREN W. POFF
JOSEPH D. GLASSMEYER
JOHN M. ATHERTON

When Mr. Wheeler died in 1962, the company's leadership passed to Charles Messer as the oldest of Frank Messer's sons. Charles had two brothers and a sister. One brother, Louis, worked with the company and its affiliates for his entire career. Neither the other brother, Sam (who had a long career as an actor under the stage name Robert Middleton), nor his sister, Sarita, worked in the company. When Frank Messer died, the company was placed into a trust with ownership to be split evenly among the siblings upon the death of their mother. As the siblings and the company matured, the trust was amended to pass ownership on to the grandchildren of Frank Messer.

Charles had been active in the Messer companies under Mr. Wheeler's leadership. When Charles succeeded to leadership in 1962, at age thirty-nine, he set about making some changes. Charles had not been included in many of the networking opportunities and relationships that Mr. Wheeler valued. Charles may have suspected that part of his exclusion was because he was Jewish, but for whatever reason he did not pursue retaining those relationships after Mr. Wheeler's death. In fact, it could be said that Charles had an honest skepticism about relationship contracting. He felt that the fees were too low (because you were always lowering fees to make the owners happy) and that the expectations were way too high (as the owners were always expecting to get something extra because the contract was awarded without competition). Charles valued lump-sum bidding, a system where the smartest survived and (he believed) the fees were highest. Over the course of the sixties, Messer exited the negotiated marketplace and concentrated on lump-sum bidding.

Charles was also concerned that there would not be a capable family member to follow him. He decided that the best way to manage that future risk was to enter the real estate development business. His reasoning was that buildings were illiquid, so it would be hard to lose all of the family wealth during the next generation. Under Charles's leadership, the Messer real estate companies—primarily Garfield Enterprises—grew steadily through both purchases and construction. Messer built and owned Park Tower on Burnet Avenue near Reading Road and Garfield Tower on Garfield Place downtown. Messer purchased the Paddock Hills Apartments at Paddock and Reading, Airy Towers Apartments off of Colerain Avenue, and many others. One of Garfield's properties was the Alcoa Building, which was developed for lease to Alcoa Aluminum and therefore contained almost exclusively aluminum metal shapes (later the offices of the Dan Beard Council of the Boy Scouts of America).

COMMUNITY ENGAGEMENT

Charles Messer was a generous man, but he seemed to have a real fear that his generosity would lead to more "asks" and that "giving" was a competition that he couldn't win. This fear caused him to be so defensive that often people felt ignored or insulted. After the Boy Scouts leased the Alcoa Building, their leadership decided that they should ask Alcoa to donate the building to the local Boy Scout council. The leaders contacted Alcoa and made a trip to Pittsburgh to make the ask . . . only to be told in Pittsburgh that Alcoa had occupied the building under a lease and was sub-leasing it to the Scouts, but that the building was owned by Charles Messer. At the time, Charles was a member of the Boy Scout Council board, but he hadn't mentioned that he owned the building, evidently out of fear that he would be asked to make the donation. He ultimately helped the Boy Scout Council own the building, but a lot of people were frustrated by the process. If today we make it a point to celebrate

Park Tower Apartments

Victory Tower

Alcoa Building

Paddock Hills Apartments

PART OF THE MESSER REAL ESTATE PORTFOLIO.

**our community investments, it is because we want to have others
feel good when we are investing our resources to do good.**

During the sixties, the construction side of Messer got steadily smaller as capital and leadership energy were concentrated in real estate development. In fact, Charles became so convinced that real estate was the future that he adopted a strategic plan to get the family out of construction, working with other company leaders to set up a successor construction company: Messer Perin Sundahl & Associates, Inc. In a glimmer of things to come, MPS was formed in 1966 as a path to employee ownership of Messer's construction operation. Jim Perin moved from Messer along with a group of "volunteers" to form the new company, which was funded partially by employee investments. The commitment was that over a period of years "Big Messer" would increase its focus on real estate development and MPS would step into the void created in the commercial construction markets.

Three things kept that plan from becoming reality. First, Charles soon realized that construction created the cash flow that allowed him to invest in real estate. Real estate could be profitable, but he was right in his assessment that it was illiquid. In fact, he was so proud of 100 percent occupancy—and set rents below the market to achieve that goal—that very few of his properties made any cash money. They might have been appreciating in value, but there was no positive cash flow. Charles needed to stay in the construction business or change his approach to development. He elected to stay in construction, creating competition for projects that assured MPS would struggle.

Second, key employee leaders recognized that the commitment to the new enterprise was not "unqualified," so they simply refused to volunteer for the assignment. This left many good employees who did agree to work at MPS out on an island without strong support. MPS became the Siberia of Messer. You could go there and you might survive, but you would not enjoy the experience.

Third, and most important, the employee investors had no real voice in the running of MPS. They put in their money based upon a promise . . . then heard nothing more until about 1980, when they were informed that their investment had not appreciated and that MPS would be closed.

MPS was a good idea and MPS had some good people, but MPS failed because leadership was not fully committed to the strategy. In the end, MPS closed and its legacy was in learning that leadership commitment is a requirement for successful change.

In some ways, Charles Messer was an innovator. In the early 1960s, he became interested in a new hoisting machine called the "French crane,"

whole lost money. In hindsight, it is clear that many opportunities were lost due to this internal competition.

INCENTIVIZING SELFISHNESS

In an effort to elevate the game of non-owners, Charles Messer and Les Sundahl instituted a metrics-based incentive system. The basics of the system were that the office leader of a project, who did the buying, would receive a certain percentage of the "buy"—the difference between the line-item cost in the estimate for an operation and the price in the subcontract for that item. The field leader of the project would receive a certain percentage of the "labor savings" on the project—the difference between the estimated labor cost of an operation and the in-place cost as performed by Messer crafts. Unfortunately, this narrow focus on metrics caused tunnel vision that often got in the way of doing a good job. "Buying up," i.e., adding money to a subcontract to lower risk or save time, would cost the employee money out of his pocket; therefore, it just didn't happen. "Buying down," i.e., taking more risk onto Messer to lower the subcontract price, added money to the employee's pocket, and it happened often. If the result of "buying down" was that Messer labor overran the estimate, the delta theory of success kicked in, and the office leaders celebrated the evident fact that they were indeed smarter than the field. The same thinking permeated the field. If labor could be saved through cutting corners or unsafe practices, it was money in the employees' pockets—even if that savings cost the company relationships with customers. An interesting omission from the definition of labor costs was the cost of workers' compensation insurance, which lagged accidents by several years. Some individuals made real money gaming the company bonus system, but in hindsight it is very clear that the company's bottom line, reputation, and opportunity flow suffered as a result of this system.

In this environment it is hard to say that Messer thrived, but the combination of good people and a diminishing appetite for construction, resulting from the focus on development, resulted in small but positive profitability in a challenging marketplace. As Charles Messer focused more of his energy on development, he needed a leader for the construction operation. Charles selected Les Sundahl, who had come to town as a young engineer with the Wells Brothers team. Les was very smart and a gifted builder; however, it was clear that he would always work for Charles, never replacing Charles as the true company leader.

Les's appointment was significant, however, in that it continued the pattern of non-family leadership at the highest level. When Les was appointed president, he was chosen over Louis Messer, who was then a vice president and leader of multiple projects. Charles could be a tough guy to work for, but the message was clear: performance would earn promotions, even over family relationships. This was one of the reasons that non-family members stayed with the company.

MANAGING CHARLES

The one person who could manage Charles Messer was Miss Shrenk. She stood up to him, talked back to him, and sometimes went her own way, with or without his knowledge. One day in the middle of one of his development projects, Charles came into Accounting and asked Miss Shrenk how much money was in the checking account. She pulled out a checking account ledger and showed him the balance of about fifty thousand dollars, and then told him he couldn't spend any of that because it was needed to cover the weekly payroll. Charles left in disappointment, and after he left Miss Shrenk pulled out another ledger and said to the people in Accounting, "Never let him see this one; it has more than a million dollars in it, and if he knows about it being there, he will spend it all."

Another example of the illusion of control—or at least the reality that both sides can play the control game.

One of the many competitions that Charles orchestrated was the battle to select the future leader of the construction operation. The two top candidates were Al Berndsen and Loren Poff, two very different people. Loren was smart, polished, a civil engineering graduate of Ohio State University, and identified with the office in the struggle between the office and the field. Loren was a very good person. He grew up in modest circumstances in Granville, Ohio, lied about his age to serve his country by entering the Navy at sixteen, cared deeply about family and friends, and had served his time in the field—having famously slept on the laborers' shed floor at a project in West Virginia because the company would not grant him any per diem when he was out of town. But Loren understood that his job as an executive was to find the errors, bring them to the attention of the performers, and to encourage them to play harder because they were way behind. Loren's brilliance and

Looking southwest toward the Albee
Theater, now the Weston Hotel
(the façade was relocated to the
Convention Center).

Looking south toward the
Sheraton Gibson Hotel (now
the site of U.S. Bank Building).

Looking northeast toward the current site of
the Aronoff Theater (Keith's Building is the
twelve-story high-rise).

Al Berndsen and
Joe Glassmeyer.
The executive in
charge and the
project manager.

FOUNTAIN SQUARE UNDERGROUND PARKING GARAGE: THE CATALYST FOR REBUILDING A CITY'S CENTER.

his articulation worked very well in the office, but it did not play very well in the field.

Al Berndsen was big, bluff, very ambitious, and not too interested in competing on articulation. Because he had worked as a carpenter, Al identified with the high performers in the field and saw it as his job to get them the tools and information they needed, even if it was against company policy. Al regularly broke the rules about information sharing, providing his project leaders with cost estimates and expense reports that allowed them to make informed decisions and to know whether they were winning or losing. Understandably, Al's approach played better in the field, and the highest performing project leaders gravitated toward Al's projects. What's interesting is that Charles seemed to actually like the fact that Al broke his rules. He liked the fun of arguing with Al, and he respected the fact that Al's projects generally made more money than Loren's made. The fact that often it was Charles's own policies that held Loren's projects back did not count for much. In the end, Al won the battle because he had the support of the best performers, and everyone, including Charles, understood the impact of those performers.

MANAGING THROUGH CONTROL

One place where information was tightly managed was in the Messer Yard, the operation that purchased and rented tools and equipment to the projects. The company invested capital in tools and equipment and rightly expected a return on that investment. However, the actual accounting was tightly guarded and project leaders were often told that they were losing money, in part because of Yard charges. One service that the Yard provided was tanks of oxygen and acetylene gas for metal cutting operations. The Yard purchased from gas suppliers and made its money from charging for delivery plus a small percentage fee. Because no one in the field ever saw the accounting, the field did not know about something called demurrage, which is the rental cost on the tanks that contained the gas. So the field managed the only cost they could see: the cost of trucking. Instead of calling the Yard to return tanks, many projects simply buried the tanks to save the trucking costs. For years Messer paid rental on tanks that had long since disappeared into backfill simply because protecting information was the highest priority.

In 1968 on my first day at Messer as a co-op, I was assigned to Estimating and led back to the "bullpen" where there was a large plan table and four telephones for communicating with subcontractors. When I arrived, there were two big men standing at that plan table screaming at each other. Al screamed, "I'm the goddamn project manager!" Charles screamed back, "I'm the goddamn owner of the company!" It went on like that for a while (causing me to wonder just what I had signed on for) until Al said, "This is what Joe thinks we should do." That statement seemed to end the discussion, causing me to wonder who Joe might be.

Which brings us to Joe Glassmeyer. Joe was born in Cincinnati in 1921. He was the third of four boys in a "good Catholic family." Joe was very smart, finishing St. Xavier High School at sixteen and earning a scholarship to pursue a career in law. However, as fate would have it, Joe's dad passed away the summer after Joe finished high school. Feeling keenly his responsibilities to his mother, Joe gave up college and took a job with Messer as a water boy on the National Distillers Project in Lockland, Ohio.

As with most life choices, it is impossible to say what Joe would have accomplished if he had completed law school. The record is clear on what he did accomplish, becoming Messer's leading project manager, growing a generation of project leaders, and providing the support and guidance that led to two individuals—Al Berndsen and Pete Strange—becoming CEO of Messer. There is no evidence that Joe ever complained about lost opportunities for himself, but there is ample evidence that he put his entire being into creating success for others. If he met you as a laborer, he made it his goal to help you become a carpenter. If you were a carpenter, he wanted you to be a foreman. And if you were a college graduate, he wanted you to lead the company. Amazingly, he had the plan and the influence to accomplish all of those purposes. Joe Glassmeyer was a magnet that kept many good people with Messer at a time when appreciation, information, and compensation were all in short supply.

Joe taught me a very simple test for leadership: "Look for someone who is causing the people around them to get better results, make better decisions, and to think longer term. When you find someone who can cause others to do those things, then you have found a leader. It doesn't matter how old they are, how many college degrees they have, or what their job title might be. People who help others get better results, make better decisions, and think longer term create 100 percent of the value in this world and about 99 percent of the fun!"

One of the many people that Joe helped grow was Jack Atherton. Jack was a mechanical engineer from the Kansas State University who had many jobs before coming to Messer. During the war, he piloted Liberty ships on test voyages to South America. After the war, he worked for several companies, both large and small. He met Messer when he was a project manager

ATHERTONISMS

A Truly Unique Voice

"You steel and concrete donkeys don't know much"

"If killing time was a crime, some of you would be in jail for life"

"You'll know when winter is over; the service repairman will return your call"

"You manage like a balloonist; you think hot air will take you a long way"

"Once a public burning is in prospect, act swiftly before the kindling can be gathered"

"Trust in God but lock your car"

"This time your tongue outran your brain"

"One advantage of being stupid is that you never get lonely"

"Never interrupt when you are being flattered"

"Good humor is the best club in any bag"

"Let's walk and talk!"

Jack Atherton and Pete Strange. A handbook engineer and a steel and concrete donkey.

for one of its subcontractors, Walterman Plumbing. Jack was a character. He believed that he knew more than everyone else, and on the subject of building systems, he may have been right about that.

He and Charles Messer met at a project for Ford Motor Company. Jack told Charles at every meeting that Messer didn't know what it was doing. Finally, Charles responded, "If you know so much, why don't you come to work for me?" Jack always claimed that he said yes just because Charlie needed the help so badly.

Jack's communication challenges were legend. On a project for P&G, he was assigned to review design drawings in progress. His approach often was to take a red marker and mark around the entire border of the drawing and write: "Everything inside this box; no good!" If a designer used NTS (Not to Scale) on a drawing, Jack would write "Not Too Sure?" next to it; or "What co-op drew this?" in the question box. He had such a knack for irritating people that Miss Shrenk would not allow him into her office . . . even to pick up his paycheck. I once took him to a value engineering meeting with an owner and design engineer where his first suggestion to the owner was: "Start by hiring a competent designer!" When I protested, Jack said, "You're just a steel and concrete donkey; you don't know much."

Joe decided to like Jack; and Joe protected Jack from Jack's own worst enemy—Jack. Because of Joe (and Jack's real talent) expertise in building systems came to be respected at Messer at a time when most general contractors simply did not become involved in their subcontractors' work. Jack's talent and Joe's leadership laid the foundation for what is today the Building Systems Group, another key differentiator for Messer.

TRUE LEADERSHIP

Around 1974, Messer had a disaster of a project at Morehead State Teachers College in Kentucky. We were constructing the new student center to be named after the long-term and still-serving college president. The building was clad with marble; however, before there was even a certificate of occupancy, the marble started separating from the building and was at risk of falling to the ground. As usual, there was a great debate as to the cause of the failure, a debate that was not helped by Charles Messer's public statement that he had known from the beginning that the design was wrong but that it was his place as a contractor simply to execute what was on the drawings. The result of some heated exchanges between the parties was a "negotiated settlement." The terms of that settlement were that the Commonwealth of Kentucky would pay $250,000 toward removing and properly re-hanging the marble, and that Messer would pay all the costs in excess of that amount. The Commonwealth, being understandably upset over this additional cost, took one more action after the settlement; they hired the meanest architectural inspector in the state to make sure that Messer's cost exceeded their own.

I was working in the office, and Joe, who by this time was vice president of Operations, stopped by my desk to invite me to go to Morehead to kick off the remediation. I politely declined, believing that my best strategy was to keep my distance from the disaster. He insisted and I responded that I should not go because, being emotional, I was likely to say something embarrassing to the SOB architect. Joe explained that he was a company vice president and I was not, so it would be in my manifest best interest to do what he asked. Now having a clear understanding of the situation, I agreed to go with him.

I will never forget that meeting—just the architect, Joe, and me in a small job trailer. Joe sat down, looked the architect in the eye and said these words: "Bill, I know why you are here. I can't do a thing about that, but I want you to know where I stand. I may never be able to make you like me, but you can never keep me from liking you." A less talented, less likeable, less competent person could never have pulled this off, but with Joe Glassmeyer it worked. Not too long after that meeting I was invited to a cookout in that architect's back yard; and he became our ally in turning a disaster into a positive outcome. Joe's positive attitude, his genuine love for others, and his competence as a problem solver changed the game—but it all started with that single bold statement. To this day Joe's statement is the single strongest declaration I have heard from a project leader.

In 1968 when I joined Messer as a co-op, about 50 percent of Messer's craft workforce was African American. Many of them had come north in the Great Migration and found work when the construction business boomed because of World War II. The African American workers were either laborers or cement finishers because they were not allowed into what were then called the "skilled trades"—although anyone who has finished concrete using wet screeds can attest to the skill required. Even in the so-called unskilled trades, a second migration of white Appalachians in the

Quality Inn, Covington, Kentucky. In 1972 one of the largest revolving restaurants in the world.

fifties and early sixties displaced most of the African Americans, so that we have had to work hard to reestablish diversity in our workforce. Messer was not better or worse than other contractors in this regard; however, given the conversation that Frank Messer had with the Post Office Department executives in 1932, Messer was perhaps more ready for change. When the country was rocked with race riots in 1966, Messer leaders stepped forward and established differentiation with regard to inclusion that remains key today.

A DEFINING MOMENT

Following the Great Migration from the South, a large percentage of Greater Cincinnati-area construction workers were African American. They would begin and end their careers as laborers or cement finishers, while their white counterparts could "move up" to higher-paying jobs as carpenters, rodbusters, and even superintendents. In 1968, Charles Messer engaged Jim Perin in a conversation beginning with the question: "Jim, how many of our carpenters and rodbusters are Negroes?" Jim replied with the obvious answer of none, to which Charles replied: "Jim, do you think that is right?" Jim's answer was that it was up to the owner of the company to decide what the company stood for and that Jim's job would be to implement those policies. Charles stepped up and changed Messer's world. He said, "Jim, I don't think it is right. Do something about it."

Jim had a lot of standing for this assignment because he had led Messer's labor relations efforts for years, at various times serving on and chairing almost every negotiating committee. Jim knew the lay of the land, so he first approached the carpenters' union business manager, who recognized that change was needed and who believed that allowing qualified African Americans into his union would be a positive step. With his agreement, Jim set up a school and recruited three African American laborers (who served as carpenter helpers and were familiar with carpentry work) to take classes so that they could apply to take the test to enter the carpenters' union. After some weeks of schooling, two of those first three candidates passed the test and became carpenters—and a barrier fell.

The rodbusters' union was much more of a challenge. At the time, the union gave a difficult written test to potential apprentices;

**Miami Fort Stack Base. A diverse workforce
placing 5,023 cubic yards of concrete in a single pour.**

so difficult that no one "passed." Entrants were then judged on subjective criteria, which meant that the status quo would be maintained. Jim—one of the most honest men I have ever met—did what he felt he had to do. He managed to get hold of that test and started another school. Given that he and another graduate engineer could barely score passing grades on the test, he did not attempt to educate his pupils on the principles. Instead, he drilled them to memorize the answers. After some weeks, two of them took the test and they scored the highest grades seen in many years, removing the opportunity to dump them for their color. Messer promptly hired those two apprentices.

But in this case, there is a "rest of the story." The rodbusters waited until Messer had a schedule crunch—a slab that had to be completed during shutdown at a GM plant—and then the entire rodbuster crew of thirteen came into the field office and informed the project superintendent that there would be no work until the two African Americans were laid off. The superintendent called Jim, the customer called Jim, and Jim went back to Charles and asked him if doing the right thing included having a contract cancelled or paying damages to the owner. With Charles's support, Jim called the project and talked to Messer's rodbuster foreman, a crusty guy named Harry Brewer. Jim asked Harry if he would go to work with the two apprentices if Jim asked him to, and Harry roared that nobody could tell him what to do; he would work if he decided to work. Jim replied that he hoped Harry and some others in the crew thought well enough of the company and of Jim to do this for him, whereupon Harry told Jim he was a bastard for bringing up loyalty and hung up on Jim.

An hour later, the project superintendent called to tell Jim that the entire crew had returned to work.

Loyalty is not a human trait; it is an employment strategy. If the company is loyal to its employees and finds ways to show that loyalty by doing right for employees, to employees, and in front of employees, then some of that loyalty might be reflected back from the employees to the employer. Harry Brewer didn't lead his crew back to work because he was afraid of Jim; he returned to work because Jim had treated him fairly over many years.

A big change during Charles's leadership came in 1966. Fearing that the non-family leaders might leave the company, he granted employment contracts to a number of executives, including Al Berndsen and Loren Poff. Among the provisions of those contracts was a commitment to pay each executive a certain percentage of his (no contracts for women, even though Miss Shrenk was assistant secretary of the corporation) annual salary for ten years after he retired. Al valued that provision highly until he realized that, as a result of the construction operation getting steadily smaller in favor of real estate development, there were almost no people younger than him working at Messer. As things stood, when Al retired there would be no one left to earn the money to pay out his contract. That aha moment led to Al's commitment to two key strategies— recruiting young people and long-term growth.

Those contracts worked well for the recipients and kept some really good leaders at Messer, but they weren't good for the company as a whole. Because there was an earnable bonus calculated based upon annual company earnings, those with contracts had a hard time thinking long term; and because only a few people got contracts, some leaders felt undervalued and left the company—not a good result from a strategy designed to keep good leaders.

Sander Hall at the University of Cincinnati.

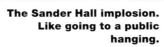

The Sander Hall implosion. Like going to a public hanging.

One Lytle Place. Hard to build after many rounds of optimistic estimating.

4

STRATEGIC PLANNING
[1975–1988]

The Players
CHARLES M. MESSER
ALFRED C. BERNDSEN
JOSEPH D. GLASSMEYER
LOREN W. POFF
WILLIAM H. DULANEY

Charles Messer's health began deteriorating in the sixties, first from a heart attack and continuing heart problems, and then with diabetes and a number of other ailments. Charles was forced to look toward a future with others leading the company. In 1975, Les Sundahl retired and Charles named Al Berndsen president, ending the competition between Al and Loren that had divided the company. In that same year, Charles named Al as successor trustee to the family trust, assuring that Al had control over decision-making if Charles and his brother, Louis, became incapacitated.

Al embraced the Messer leadership position with real enthusiasm, including raising eyebrows by ordering a new Cadillac on the day he was promoted. Most important, Al set about the business of bringing construction back to the forefront of company priorities. Al's case was simple: construction provided far higher cash flow per dollar of invested capital; therefore, construction should lead, and real estate investment should only be considered with distributed capital. It took a few years for Al to fully convince Charles; however, in 1978, the company began moving down the long path of divesting real estate and focusing capital on growing its construction operations.

Al's goal was growth. He believed that both he and the company deserved greater stature in the industry and in the community. To achieve Al's plan, the company increased hiring and focused upon doing more work outside Cincinnati. This was not a return to traveling across the country with customers as in the fifties and early sixties, but a greater focus on bidding work in central Kentucky, central Ohio, and West Virginia. These were places where Messer had regularly worked over the years, but now the goal was clear: grow revenue!

While Joe Glassmeyer was the operations leader, Bill Dulaney was the chief prophet of growth. Bill had joined Messer as a project engineer after attending Morehead College and quickly rose from the field to the office, becoming an officer after Al became president. Bill was a dealmaker. He would wheel and deal with designers and owners to get the award of the project, and then he would wheel and deal with subcontractors and suppliers to buy the project. Very often the project team was left to fill the gaps created by Bill's deals. Bill saw distance as an appropriate buffer between his deals and the discipline of the company; so Bill became the executive leader for many of the company's out-of-town projects. Bill was a very smart guy and could usually find a solution when his fast and loose dealing resulted in problems. Unfortunately, he modeled some approaches to risk management that, in the hands of those with lesser skill, created real problems. Bill was very good at identifying situations where subcontractors had overpriced a risk and would delete that work from their scope and generally find a way to perform it for a profit. There were several cases when others tried to follow this path that resulted in huge losses—either because the subcontractors misled them or because the buyer simply didn't understand the risk.

By the early eighties, Al's growth target for Messer was: "To be in the Top Ten on the *Engineering News-Record* list of the 400 Largest Contractors." At the time, Messer was about two-hundredth on that list, so achieving this goal required some real growth. Unfortunately, during this period the goal of selling the job often took precedence over margin and risk management; so after a nearly disastrous year in 1984, the company returned to more moderate growth strategies.

CELEBRATING SALES

As the company worked to increase sales, Messer became involved in a number of developer-led projects. Most of these projects were successes, following well-established models for HUD-sponsored housing; however, one project—One Lytle Place—was a real challenge. The project itself was not unique—a high-rise apartment building in Cincinnati. What was unique at the time was the targeted market—upscale renters—requiring much higher levels of amenities and finishes. Messer estimated this project many times over four years, each time desperately trying to close the sale. In hindsight, each estimate contained new, positive assumptions intended to lower the price while accepting new risks. When the celebration of the sale occurred four years after the first estimate, the aggregation of all of those positive assumptions assured that the project would lose money. We lost sight of the main goal—a successful project—to serve a short-term goal—a successful sale.

One great thing came from Al's desire to grow Messer. Because he knew that the company could not achieve his goals with its current strategies, Al initiated a new (for the company) strategic planning process. In 1982, a group of twelve people were invited to a two-day planning meeting in Lexington, Kentucky. This planning session was different from anything that Messer had done before. Young voices were included in the conversation, the process used professional facilitators, and non-operations people (e.g., accounting) were allowed to participate—probably because it was easier to include them than to deal with Kathy Daly if she was excluded.

Al started with the premise that big projects were being awarded to "construction managers" (that new definition for construction services that implied a more professional approach) and since Messer did primarily traditional lump-sum work, we were not in the game. The big question became: What do we need to do to balance our sales to 50 percent lump sum, 50 percent negotiated?

The answers to that question changed Messer's focus, changed Messer's processes, and ultimately changed Messer's culture. Throughout our history, Messer had been focused upon project needs, project goals, and project outcomes. We all knew in our hearts that we were better builders than our competitors. We just wrote off the jobs that were negotiated with others as the result of slick salesmen and uninformed owners. Now, we didn't have the choice of writing off those projects; we had to focus upon a new group of project participants—we had to focus upon customers.

CUSTOMERS COUNT

In the seventies, Messer bid on a new parking garage in Cincinnati at the corner of Seventh and Vine for Federated Department Stores (now Macy's). For decades, Messer was Federated's builder, doing projects all over the country. That relationship weakened after Mr. Wheeler's death, which was fine by Charles Messer because he preferred lump-sum work; however, Charles Messer and Fred Lazarus, who led Federated, knew each other well. Despite the fact that Charles preferred lump-sum work, like most contractors, he did not like the open competition that came with that contract model. He bid the Federated garage expecting that Messer would be selected based upon our long-standing relationship—even if we were not low bidder. Charles was aggrieved when Messer was not low bidder and the contract was awarded to another general contractor, and Charles made his displeasure clear to Fred Lazarus.

A few years later, when Federated made the decision to build a new corporate headquarters, Fred sought to mend the hurt by inviting Messer to help plan the project. Charles, who was still angered over not having built the garage, told Fred when he called that Fred's previous bad decision had put the project beyond help—expecting both an apology and another plea for help. Instead, after Charles told Fred that he couldn't save him from the bad design on the already-built garage, Fred simply went out and hired another contractor to build the corporate office.

All too often during that period, we spent time and energy trying to convince owners that we were smarter than they were, rather than convincing them that we were ready, willing, and able to help them.

As a result of strategic planning, we started conversations about company structure, project structure, and, most important, employee growth. Equally important, we got to know each other at a much deeper level. We formed committees to get the "strategic-planning work" done, which allowed individuals to show their leadership abilities and understand the company in a much deeper way. Not every idea found traction, and not every employee embraced the change, but the foundation had been laid for a new way of interacting with each other as leaders and as employees.

THE STOPPER IN THE BOTTLE

Two of the participants in that first strategic planning session were Jim Perin and Steve Messer. In 1982, with Messer Perin & Sandahl closed, Jim Perin had replaced Dick Glassmeyer as secretary of the company. Even though he was nearing seventy, Jim's sense of loyalty was such that he could not bring up retirement; however, it was clear that the Messer family had a replacement if and when he did retire. Steve Messer, one of Louis's three sons, came to the company with both an accounting and a law degree. While Steve was assigned to Operations—because all leaders came from Operations—his path to leadership would be through Finance and Administration. This was obvious to all, including to Kathy Daly, who had no path to leadership other than Finance and Administration. Kathy could see the writing on the wall, so she went about the business of networking toward new employment opportunities. Kathy joined committees at the Chamber of Commerce, and she sought board membership at nonprofits, all for the purpose of creating options for her future. The good news for Kathy was that one day Steve Messer, in a real example of courage, went to his family and told them that their plan wasn't right for him. He wanted to have a career that he defined, not one that was defined for him. When he left the company, a path was cleared for Kathy. The better news was that while seeking options, Kathy had modeled the path to community engagement, which is still a core strategy at Messer.

After that first strategic planning meeting, Messer had a new set of goals. Growth was at the top of the list—we had yet to experience the harsh side of growth as a leading metric—but we also had some other goals in support of growth. Those goals included:

- Balancing sales to 50 percent lump sum and 50 percent negotiated work.
- Growth of each employee through a structured education process.
- Hiring more graduate engineers to elevate the project manager position.
- Adopting "leading edge" project management tools.

THE PROJECT MANAGER POSITION: ONE MORE JOE STORY

One of the facilitators helping with strategic planning asked each of us to use an analogy to describe the job of the project manager (the person responsible for all of the construction site activities). As you might expect, many military analogies were used, from the sergeant major to the general. Joe, as usual, stood out. Joe described a project team as a symphony orchestra—each player having just his or her part of the music, with the conductor (the project manager) being the only person with a complete score (the plan). When everyone did their part with skill and commitment, and the conductor led with

clarity and passion, the team created something much larger and greater than any one player could create alone.

Many years later as board chair of the Cincinnati Symphony Orchestra, I was asked what attracted me to classical music. I responded that the CSO is the perfect analogy for what Messer does when we are at our best. We bring together people around a plan, and if we can get them to do their part with skill and commitment, and if we lead with clarity and passion, we will create something great.

We had goals and we had committees, but what we did not have was a management structure to support the change we desired. Because of the competition that Charles Messer had orchestrated (pun intended) between Al Berndsen and Loren Poff, the winner tended to see himself as king, with everyone else either a loyal subject to be guided or a loser to be controlled. This was pretty much the model for most construction operations—there were benevolent dictatorships and there were harsh dictatorships—but mostly there was just one king and everyone else reported to the king and did his (not ever her) bidding. So, Messer had a challenge: every new idea had to come to Al for his input and permission. Not only did this slow everything down, but it sapped the energy of those doing the work because they constantly questioned whether their ideas and effort really counted, or whether Al was just going to do whatever he had already decided in his own mind.

At this point, Al's devotion to growth served us well. One of Al's motivations toward growth was a vision of executive leadership that was at a new level. He saw himself as a leader who should direct, motivate, and reward, but not get caught up in the day-to-day work. Al had actually adopted this vision early in his career as an executive. The good side of this vision was that Al would allow those he trusted great latitude to act; the down side was that he created that latitude by not showing up for work until 10:00 a.m., and he spent lots of time "giving us space" by playing golf. With a commitment to growth and with lots of extracurricular planning going on, Al faced a new reality: if he wanted everyone to report directly to him, he would need to spend many more hours at work to create contact time. With a little nudging from some younger folks who aspired to greater responsibility, Al agreed to engage a consultant to do an organizational analysis. Touche Ross was at that time very engaged in selling services to the construction industry and did a lot of presentations at the Associated General Contractors of America meetings. Al had come to respect their expertise, so they were hired.

BEING THE BOSS

Every time Al went to a seminar, those of us who were left behind were treated as if we should have been thinking about the topic as a natural consequence of doing our jobs. I will never forget Al walking into my office after he attended a seminar on overhead control and almost yelling, "Do you know what the 'nut' for this company is? Do you have a plan for covering the nut? Are you prepared to lay off your friends to match the nut to sales?" The answer to all of those questions was "no," which Al interpreted as the real reason why he was in charge. Opening the opportunities to attend seminars and trade meetings to leaders across the company made everyone smarter—not just the Boss.

Touche Ross sent two accountants to interview the company leadership and to study the company structure. They were really good guys who understood that they were there to help with needed change, not just to write a paper. In the end, they presented a report to Al that said Messer needed more people sharing in leadership. Because this answer met Al's goal of not spending ever more hours doing his job, and because Al had at least one person (Joe) he could trust to lead, a new organizational structure was adopted. The existing officers all got new titles as senior vice presidents; two new vice presidents were appointed (Pete Strange and Ed Miller); and Joe was put over everybody as executive vice president.

Several people received new titles, but no new contracts were signed. Those who had been officers in 1966 got the contracts; those elected to officer after that date did not. The most generous explanation for this is that Al assumed that his leadership and the fact that he had Charles Messer's trust were proof that Al could protect other non-family leaders. The least generous is that any added contracts would have diluted the pot available for those currently sharing in company profits. The reality is probably a combination of inertia (leaving well enough alone) and a short planning horizon (making this a good year for those who share in the profits rather than paying attention to the future). One result of this lapse was the loss of a key leader, Bill Dulaney. Bill was a real asset and a strong voice, both inside and outside the company, but Bill was not shy about his desire to be on equal footing with Al, Loren, and others. To Bill the key element of that equality was an employment contract that guaranteed him a cut of current profits and that gave him a retirement income. Another company offered Bill a promotion and a contract . . . and Bill left.

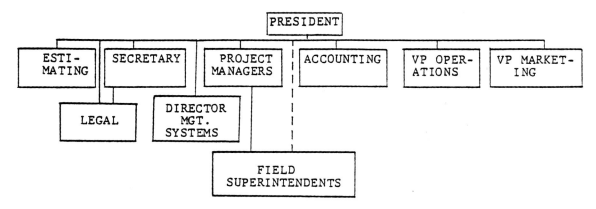

The King and his court organization structure.

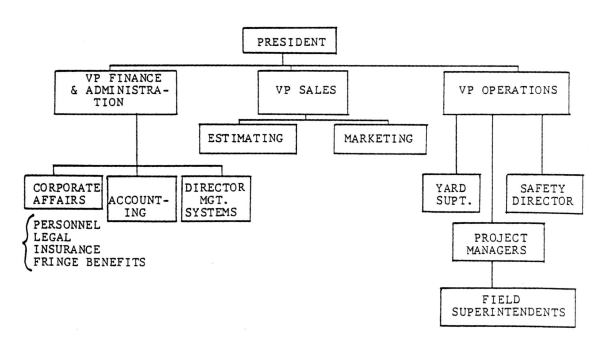

The Touche Ross proposal. The first step toward shared responsibility.

Bill's departure was not mourned by Al or Joe because they did not find much comfort in his often flamboyant approach to selling projects and buying subcontracts. Bill had also been a guiding force in Messer becoming one of the biggest farmers in Clermont County, Ohio, which was a real disaster.

In the mid-seventies, the Ford Motor Company announced that they would construct a major transmission-assembly facility in Clermont County. This looked like good deal for Messer because Ford generally awarded their work on a lump-sum basis, and Messer had performed for them in the past. As happened not infrequently, Bill Dulaney had an idea about how we could make this the "big win." Through some connections of Joe's, Bill had gotten to know two politically connected brothers who were the development kings of Clermont County. Before long, Bill and the brothers had cooked up a deal to leverage the new Ford plant by developing a thousand acres of land across the road from the new plant site. Bill sold Al on the idea, and Messer formed a partnership with the brothers. The brothers put into the partnership the thousand acres of land (along with a $3.8 million mortgage), and Messer invested cash to cover development costs. After almost two years of planning and the investment of several hundred thousand dollars from Messer, it became obvious that the development was going nowhere. The Ford plant turned out to be half the size initially announced, and there was no way to get needed utility services to our site. The brothers had a simple proposal to resolve the situation. The partnership was a separate legal entity: put it into bankruptcy.

Al was outraged at this casual suggestion, as well as properly concerned about the damage that such an outcome would cause to Messer's reputation. Al had an alternative solution: force the brothers out of the partnership and take control of the assets for Messer. The flaw in Al's plan was the assumption that the assets we would take control of actually had enough value to cover the debt. There was an appraisal prior to forming the partnership that concluded that the land was worth over $6 million, far more than the $3.8 million mortgage; but that appraisal turned out to be both optimistic and deficient. Messer got the land and the debt just in time for runaway inflation to set in and for the floating rate on the mortgage to go up to 15 percent per year. This was a disastrous outcome, which haunted the company for nearly two decades. Bill carried the blame for this disaster away with him when he changed jobs. Lots of people were enthusiastic about this opportunity on the front end, but Bill ended up owning it after he abandoned ship—so much so that the leaders with contracts were able to get their contracts modified so that the interest payments on the land did not reduce their annual bonuses.

The real impact of Bill's departure came when he began hiring Messer people away. His efforts were hit or miss because he hired whoever would

FOR SALE   1,000 ACRES
SR 276, Batavia, Ohio
CLERMONT COUNTY

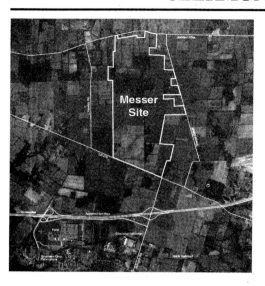

Water, Gas, Electric at street

Sewer - 1 mile away at the Ford Plant

Zoning
Industrial	518 acres
Commercial B1	112 acres
Commercial B2	28 acres
Agricultural A1/ Residential	342 acres

Terrain is extremely flat. Extensive soil investigations and environmental tests have been completed for the property and are available.

Site has access from SR32 (Appalachian Hwy.), located ½ mile north from SR32 & Half-Acre Road. Site fronts on SR276 with access to Jackson Pike and Sharps Cutoff. Property is located within one mile of Cincinnati Milacron, Ford and Southern Ohio Fabricators.

Parcel Numbers:
52-35-08F-013
52-35-08F-018
15-23-10J-021
14-23-10J-021
15-23-10F-022
15-23-10I-023
15-23-10H-025
15-23-10G-027
15-23-10G-028
14-23-10D-087
14-23-10K-088

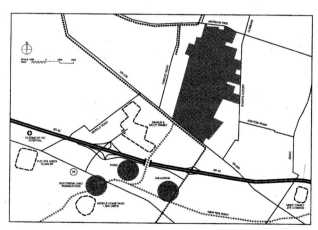

Contact: Sandy Meyer, Messer Construction Co., (513) 242-1541

The Clermont County land. A big deal that became a big problem.

come rather than hiring the best people. However, in a company where most of the leaders had been employed for decades, suddenly there was a new employment option as Bill extended a standing offer for anyone who might be interested. That meant that anyone who didn't see the progress he or she expected at Messer could get a big raise and generally a promotion, at least in job title, by moving. This new reality created a new and more practical understanding of the word loyalty on all sides. "Loyalty is a business strategy, not a human trait!"

FAILING JOE

There is not a good explanation for why Joe Glassmeyer, who meant so much to the company and to Al's success, was not given a contract. Perhaps in Joe's case, loyalty was a trait. Like every Messer employee, Joe had a small defined-benefit retirement plan ($250 per month for life) until about 1972. In 1968, Congress enacted ERISA, which regulated retirement plans. Al and Charles were extremely bothered by ERISA, which they saw as government interference in the company's relationship with its employees. At the same time Congress introduced a new concept, Individual Retirement Accounts, which would allow employees to self-manage their retirement. The combination of discomfort with the new rules and having an alternative caused the company to take action to terminate its retirement plan. Each employee participant was given the funded portion of the original commitment and was promised a check for the company's portion of continued funding as a basis for setting up an IRA.

The result for longtime employees like Joe was a five-figure initial check and an annual check in the amount of $365—which stopped in 1985. Under the contracts that others had, each contract retiree received 50 percent of salary for ten years after retirement. We tried to cure this oversight by putting in place a consulting contract for Joe when he retired, but frankly, both Al and I should have done more.

To finish the story about retirement, the departures from Messer sent a clear message that more needed to be done. One of the committees formed in the early eighties was the Profit Sharing Committee, made up of Jim Perin, Pete Strange, Ed Miller, and Steve Messer. The committee's charge was to investigate what Messer could do to create a more competitive retirement plan. This committee worked hard and concluded after some study and learning that the best path for Messer would be to consider an Employee Stock Ownership Plan. An ESOP was a new concept for all of the committee members, but all of them, including Steve Messer, agreed that an ESOP was the best path. However, after some consultation with Al, we all agreed that getting something started was more important than getting something perfect. The recommendation that was reported out was a "Defined Contribution Retirement Plan" that would allow the company board to set aside a percentage of the annual profits for employee retirement. The percentage target was 10 percent, although that was not binding for the board. The committee also recommended, and the company adopted, language that would allow the plan trustees to convert the Defined Benefit Plan into an ESOP, if and when that action might become appropriate. The learning we gained in creating this profit sharing plan and the resources that began accumulating inside the plan would be critical to Messer's future.

Shortly after the Profit Sharing Committee completed its report, Steve Messer informed his father, Louis, that he would be leaving Messer. This was a shock to everyone because Steve was clearly the family favorite for leadership, and by intellect and education, he was very prepared to lead the company. He just did not enjoy the work; and just as important, he didn't like the idea of competing with friends for a job that he didn't want. Steve went on to success in a variety of business ventures, having acted with unique courage and conviction in abdicating a throne that was his by birth.

Steve's departure opened up a clear opportunity for Kathy Daly and changed the conversation about succession. The idea of non-family leadership was obviously not new, but until Steve's exit, the odds were against it happening in this generation. There was no immediate impact because the aspirants were younger and because Al Berndsen could not contemplate leaving the job of being king, which he truly loved.

Messer's new strategic focus, its deeper level of engagement in growing employees, and its commitment to performance soon began to produce results. In 1982, the company achieved a breakthrough by winning the contract for the new Mercy Health System hospital in Anderson Township. This both marked the reentry into health care work after an absence of two decades and created a platform of learning and experience to grow a new generation of health care builders. The focus on negotiated work began to gain momentum with the confidence we felt after making this sale.

Even though we had embraced the new strategy and were finding some successes, we were still in a reactive mode with regard to customers and projects. Our sales strategy was to review Dodge Reports (F.W. Dodge, a division of McGraw-Hill, was the leading construction-industry reporting

Mercy Hospital, Anderson Township, Ohio. The first health-care construction management project.

service for our region) and to "follow up" when we found what looked like a good opportunity. Too often we made the follow-up call to learn that a construction manager had already been selected. Messer needed a proactive sales strategy, but Al and other leaders were not at all enthusiastic about hiring a salesman. While we recognized that we lost projects due to a lack of sales talent, we had the strong feeling that the sales process as practiced by the big construction managers was generally unwholesome and that sales promises were the antithesis of performance.

Those feelings had been fed by Les Sundahl's one attempt to hire a salesman. The guy Les hired looked like he came from central casting, with alligator shoes, a cigarette holder, and a smooth line about how much smarter and more urbane he was than us rubes. The one job he almost succeeded in selling was of a size and in a location well outside our current scope (we missed the award by a penny a square foot of lease cost, which probably saved Messer from going out of business). Shortly after that failed proposal, our sales hero's past caught up with him, and we learned that he had left a trail of tears across a number of projects in several states. We had a party when Les fired him.

That bad experience did not change the reality that we needed help. Luckily for Messer, the solution was already in place. As I have said several times, Al Berndsen and Loren Poff competed for the CEO position. Loren lost and was sent to the Messer workhouse: doomed to return to being a project executive with a great big title. (Quitting wasn't an option for Loren; his contract was just too good, and we were starting to create some real profits for him to share in.) So, when Al felt the need to do something immediate to elevate sales, he gave the assignment to Loren. This put a senior person on the matter, and, much to Al's satisfaction, it put an even greater focus on Loren's failure since it moved Loren out of Operations in a company that had traditionally valued Operations above all else. Whatever Al's motivation, it was a great move. Loren had the polish, the experience, the communication skills, and, most important, the stamina to sell. Loren created from scratch a deep and effective sales process for Messer. He may have lost the competition for the top job, but he was willing to put that aside and help us create a better future.

AL'S MOTIVATION

By the time Al gave Loren the sales assignment, the conversation about sales had been going on for a good, long while. What finally caused the action? Probably aggravation, as usual. Engaging employees gives them a voice, and some of those employees will choose to use that voice. In this case, one of the "young guys" invited my wife and me over to his house after the Messer Christmas party. Putting aside the question of whether it was good judgment to leave one party at 11:00 p.m. to go to another, I arrived at Tom Keckeis's house to find myself surrounded by a group of "young guys" who had a message to deliver. So, at about 3:30 a.m., while our wives slept uncomfortably on the living room floor, I was treated to prepared remarks, with bells and flourishes provided by alcohol. As the spokesman, Tom shared that, while they thought I was a pretty good guy, they really wondered about my intelligence and my sanity with regard to Messer's commitment to growth. Even Tom's dad's little company (at the time Larry Keckeis led Honnert, a small but well-respected general contractor) could see the need for a dedicated sales person; why were we so stupid as to keep putting it off? It was a good speech, and either the strength of the argument or the effect of the liquor caused me to go to the office the next morning, on Sunday, and to write about thirty pages on a yellow legal pad capturing the discussion and making the argument for immediate action. I left the writing on Al's desk and went home happy that I had transferred the problem.

On Monday, about 11 a.m. (after Al arrived and read my report), he called me into his office to ask if what I had written was a job application. I had never considered that as a possible outcome or I would have been more careful with what I wrote. After taking a few moments to catch my breath, I responded with the Messer mantra, "Operations is the path to the top; I want to stay in Operations." Al said that he was just checking and that he agreed with me about Operations; then he said he knew just who to give this to, a guy who had failed in Operations. I have often wondered whether I would have been up to the challenge. Loren may have been selected for the wrong reasons, but I am convinced that he was the right leader for the job.

Loren got us organized around sales, and he put himself on the road to find projects. He was absolutely dogged in his determination to put Messer in front of owners and to earn their work. He called on the CEO of one major hospital twelve times before he actually met the man; and Messer ended up with the project award. Loren could be a tough guy to work with, and he could be a very aggravating boss, but he loved Messer, and, in the end, he created a legacy through sales.

Middletown Regional Hospital. The project Loren sold after twelve unsuccessful visits.

PERFECT TIMING

As things came together for Messer, it became clear that we would need to earn awards on bigger projects if we were to grow annual revenues above $100 million per year. Al once again engaged Touche Ross to help us with some consulting. Touche Ross's construction guru came down from Minneapolis and started our meeting with something of a lecture. He said that he doubted that we could grow much until we broke the "$30 million dollar project barrier" and that without at least one $30 million dollar project in the backlog, a contractor simply could not sustain such high revenue levels. He asked how many proposals we currently had active for $30 million dollar projects. We had to reply "just one," but at the moment when we shared that sad news the door of the conference room burst open, and a secretary said there was an urgent telephone call for Bernie Suer. When Bernie came back into the room, he shared that we had been selected by Christ Hospital as their construction manager on the $32 million Courtyard Project—and we were on our way.

The year of 1984 was tough as we digested the lessons of growth. We had taken too much work based upon our needs and desires, without making sure there was a match with our resources and abilities. We finished the year "break-even" but only after some very creative analysis of the value of some capital assets. The new strategic focus made the future look much brighter; however, a real challenge arose when Messer's bonding company (the large insurance company that provided our surety bonds—the third-party guarantee of completion required on all public projects) began to suggest that they might not support Messer's growth. Surety had been an issue for Messer, as it is for most contractors. Maintaining the balance between the assets on the balance sheet and the book of uncompleted work is a challenge, especially if leadership is focused on growth. In the early seventies, when Charles Messer directed an ever-greater portion of working capital into real estate investments, Messer's long-time surety, Transamerica, had dropped the account. With some good work by Messer's surety agent, the account had been placed at United States Fidelity and Guaranty Corporation (USF&G). As with many surety accounts, the underwriting was based partly (obviously in this case the smaller part) upon financial strength and partly upon the performance record of company leadership—in this case, in the person of Al Berndsen. Even though the company had rebuilt its balance sheet after the real estate investments, the tight year in 1984 combined with the heavily leveraged land in Clermont County strained Messer's surety credit; and then came The Beach.

71

Christ Hospital Courtyard Project. The answer to the $30 million question.

The Beach Waterpark was one of those "once in a lifetime" projects (there just are not going to be a lot of water slides built in places where the number of days over eighty degrees does not account for one-third of the year), so it was both hard to plan and hard to execute. If it did not open by June 1, 1984, the pro forma simply would not work. So, Messer pulled out all the stops and, with the support of a trusting, committed owner, completed the project on time. But the extra effort caused the project to go well over budget; and the pro forma that would not work after June 1 certainly didn't work with 20 percent more debt burden, so there was a real problem. The owner understood the cost overruns and was appreciative of the performance (we have done many successful projects for him over the years); however, there was simply no money available to close the gap.

Messer took the only route available and provided a second mortgage. That closed the project gap but, combined with the other real estate on the balance sheet, it left Messer with very little liquid capital that could be used to address future project challenges. This was a dilemma for us and our surety. Both sides knew and respected the leadership aspect of the relationship, but the surety leaders needed a stronger case to take to their senior management.

Getting the information needed to underwrite an established but weak contractor could be a real challenge for a surety manager. The contractor would know that the default position would most likely be to continue to say yes to bonds, so leadership would simply delay sharing audited financials or work-in-progress reports until there was good news or until the data was too

The Beach Waterpark. One of a kind.

stale to act upon. In Messer's case, this challenge was increased by a well-established, well-connected surety agent who did not hesitate to go around the regional surety manager directly to surety senior management. Al and the agent would talk about providing information, but were very reluctant to act.

Everything changed in the autumn of 1984 when the surety set a deadline for getting good data. Al wanted to continue to stonewall, but there was no way that the deadline could pass and Messer remain bondable—and if Messer couldn't get bonds, Messer couldn't get work. Jim Perin stepped up to break the deadlock. Jim had enough influence with Al to get him to act, and he had the experience and intellect to drill down to the essence of the problem. Jim, with a little help from Pete, created the risk-assessment-based Work in Progress Report that we still use to communicate to our sureties. It didn't cure our balance sheet, but it did allow us to remain bondable; and it opened the communication channels we needed to maintain bonding when we employees bought the company.

SURETY COMMUNICATION: OLD AND NEW

My first visit to the home office of USF&G was in the spring of 1985. As the new Operations VP, I joined Al on this trip. With the report that Jim and I had created in hand, I was full of energy and pride, believing that we had something new to share and that this meeting would be the foundation of a deeper relationship.

Al, our surety agent, and I were ushered into a sixteenth-floor conference room at USF&G in Baltimore. Attending from USF&G were our local bond manager, a regional underwriter, two vice presidents, and the president of USF&G surety. All of them except the underwriter had known Al for years; nobody knew me.

Not knowing the order of business, I waited expectantly for someone to ask me what I had to share. Instead (as I would find out later the agenda was set by the difficulty the underwriter perceived in getting information from Messer), the underwriter suggested as a first order of business that, in light of the leverage to our balance sheet created by the Clermont County land and the Beach Waterpark second mortgage, Messer agree to provide audited statements every quarter so that the surety could have reliable information on a regular basis. I thought this was a little over the top but certainly understood the concern.

Al reacted as if he had just been accused of grand theft. He leapt to his feet, knocking over his chair, and shouted that he wasn't going to be insulted by some little twerp. Then he started screaming, mostly incoherently, about his lifetime of accomplishments, his unimpeachable integrity, and the low character and bad habits of underwriters. I figured we would be ushered to the door and would spend the return trip talking about how to keep the company in business.

Instead, after several failed attempts to get Al off the ceiling, the USF&G president and a vice president each took one of Al's arms and led him gently out of the room. I asked those left in the room the rhetorical question, "What do you normally do now?" but received no answer, so we sat in silence.

After a while the two USF&G execs ushered Al back into the room, and the president turned to the underwriter and said, "Ed, I'll ask you not to speak again in this meeting." When we finally did get to my report, my information was well received, and I am very proud of the trust we have earned from our sureties; but I have often wondered whether Al's approach would have carried the day without my input. I could see that I had a lot to learn about being an executive.

Surety challenges notwithstanding, by 1988 Messer was on a path to steady growth and was creating record profits. Clearly, when Charles Messer picked Al Berndsen to run the company, it was a good decision. Charles also named Al to follow Charles's brother Louis Messer as trustee of the Messer family trust that contained 92 percent of the company stock. In fact, Charles had vested Al with so much personal power that when Louis Messer and two of his sons asked Al to attend a meeting with their attorney to discuss their buying out the rest of the Messer family, Al not only killed their plans on the spot, he left the attorney's office with a greater share of the profits than was in his original contract. Al was running the company, but Al was sixty-five years old; and even if he intended to work forever, there was no evidence that God would support that commitment with a record lifespan.

Louis Messer died soon thereafter, and his death was followed shortly by the death of his sister Sarita. Messer was now 92 percent owned by fourteen grandchildren of Frank Messer, only one of whom, Frank Messer, Jr., was an employee of the enterprise. Within weeks of the passing of the second generation, Al and the "Cousins," as we called them, developed a special relationship. I named it the "evil stepfather/idiot stepchildren relationship." If the Cousins asked a question, the response was they were too stupid to understand construction; and any information that Al volunteered was taken to be a lie, just because it came from Al. And Al loved it.

Al was a born fighter; he loved the fight and he loved the fact that he was in charge. The trust document was clear: Al could vote 92 percent of the shares as he saw fit. The Cousins might sue him, but it would be after the action was taken. They had little chance of stopping him from acting. This was all a game for Al, but it was deadly serious for the younger generation of Messer employees. We still had thirty years to work. Who would we work for; what level of commitment would there be to the growth plan; what sort of opportunity would there be for us if a new generation of Messer family members joined the company; and what was Al's end game: did he just want to enjoy the fight, or did he have a plan for the future that we could support?

As the longest tenured of the younger generation, I got the assignment of bringing these questions to Al. I probably could have framed the question better, since his first response was to tell me that his plans were none of my business and that I should just trust him. My response to that was something like, "I need to go now so I can fill out some job applications," which didn't help the situation. At any rate, after we got done yelling at each other, we did have a good conversation about the choices that I and others would have to make about our futures, and we agreed that those choices should be based, to the extent possible, upon real information not conjecture. With a little bit of help from Jim Perin, over the next week or so we were able to convince Al to call a shareholders' meeting.

One of the challenges Al faced was that the Cousins had even less information than the employees upon which to base a decision about the future. As best we could tell, what they knew about Messer amounted to hearing statements at family gatherings from their dad or uncle along the lines that the Messer family was wealthy. Even Frank, Jr., who worked for the company as a project executive, was not in a position to see the financial statements. Adding to the challenge, the family was not close, so there was broad concern that someone among the group might be getting an advantage. In his typical way, Al decided to take this problem head on. He would hold a meeting and he would educate the stupid Cousins as to the reality of Messer. That meeting was held on May 7, 1988, and it began the new era at Messer.

I may have been the biggest beneficiary of the shareholders' meeting. In preparation for the meeting, I studied the details of the company's performance, starting in 1932 and coming all the way up to 1987. Never had an Operations leader from Messer been allowed to spend so much time studying

FRANK MESSER and SONS, INC.

SPECIAL SHAREHOLDERS MEETING

MAY 7, 1988

Company Offices, 4612 Paddock Road
Cincinnati, Ohio

AGENDA

OPENING STATEMENTS   Alfred C. Berndsen
  Purpose of Meeting
  Messer/Past-Present-Future

THE EARLY YEARS   1932 - 1962   James C. Perin
  The Leaders
  Business Philosophies
  Significant Events
  Results

THE MIDDLE YEARS   1963 - 1979   Loren W. Poff
  The Leaders
  Business Philosophies
  Significant Events
  Results

THE RECENT YEARS   1980 - 1987   Alfred C. Berndsen
  The Leaders
  Business Philosophies
  Significant Events
  Results

_____ O _____
BREAK
_____ O _____

THE FUTURE   Alfred C. Berndsen
  Extend Present Course
  Merger/Acquisition

OPPORTUNITY   Peter S. Strange
  ESOP

SUMMARY   Alfred C. Berndsen

QUESTION PERIOD

_____ O _____
ADJOURNMENT

The agenda for the Messer shareholders' meeting.

finances. Even better, I got to review those numbers with Jim Perin, who had been there since 1936, so he could connect most of the results with the decisions that drove them. I learned the impact of capital allocation, the details of compensation outcomes, and the harsh realities of a high-volume, low-margin business. It changed how I felt about risk and reward, and it changed my understanding of the long-term nature of decisions.

In addition to that education, I got the opportunity to speak at the shareholders' meeting about one possibility for the future—employee ownership. In preparation for the meeting I really dug into the subject so that I could present with conviction. After all, this was my opportunity to have a say in the future of the company I worked for. Along the way, I began to believe that employee ownership might be a good idea. I am honestly unsure as to how seriously Al thought that option might be received; I think his main goal for the meeting was to scare the Cousins into leaving him alone to run the company. If that was his goal, it didn't work.

Al began the meeting by introducing everyone and describing their interests. There were thirty-one shareholders in attendance—the fourteen cousins and a number of current and former employees who, for one reason or another, had either received shares as compensation or had been allowed to purchase shares. Jim Perin presented the history, detailing major decisions and showing the impact those decisions had upon the balance sheet. Al shared the results under his leadership, which were generally positive, and then he presented what he saw as the four options for the future:

- Sell the assets, which would yield a small percentage of the book value of the enterprise.

- Sell to a competitor, which Al could not recommend because once it became known that Messer was for sale, the competitors would probably just make offers to all of the key employees, so that by the time the negotiations took place there would be nothing left to sell but the assets.

- Stay the course, which meant that the owners would forgo access to their capital and even give up dividends so that Al could use that capital to continue to grow the company.

With those three cheery prospects on the table, Al introduced me to present "one more option." I presented the combination of tax benefits and potential of an ESOP with what I thought was great enthusiasm, but did not get much response.

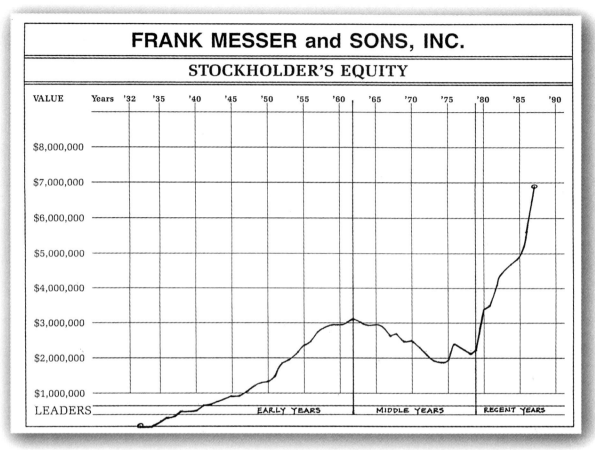

FRANK MESSER and SONS, INC.

STOCKHOLDER'S EQUITY

Stockholders' Equity. The chart that made the "Cousins" mad.

When the time came for questions, the reason for the Cousins' lack of enthusiasm became obvious: I have never seen fourteen madder people in my life. They didn't know much about Messer, but they did know one thing … they were a hell of a lot richer than these numbers showed. Someone had stolen their money. And on that note the first open shareholders' meeting of Messer ended.

Louis Kelso

5

NEGOTIATIONS
[1988–1989]

The Players
ALFRED C. BERNDSEN
GUY M. HILD
THOMAS D. HEEKIN
PETER S. STRANGE

First, I should share that I did not naturally support the concept of employee own-ership. Like many others in Messer, I had spent my entire life in construction. I knew firsthand that the industry treats human beings like a commodity—hire 'em when you need 'em, fire 'em when the job is done. The offsetting positive of this difficult employment environment was that we as employees had total freedom from loyalty. If at any moment the Messer leadership did something that offended me, I was totally free to pick up my tools and go to work for the company across the street. I basked in this freedom because it fit with my notion of employment. I believed what my father—who quit a job every six months just to prove he could— had taught me. That is that employment is the worst deal you can make. The boss gives you compensation, which is basically money, a renewable resource that can be replaced a thousand ways. On the employee's side of the contract, you give up 100 percent irreplaceable time. If the boss wastes your time, there is not a court in the land where you can go to sue and get it back (this part I added after I took contract law in college). There is a fundamental of contract law called the "rule of equity" which says that the party with an extreme advantage in a contracting situ-ation should expect to have the contract construed against them.

Given my attitude about employment, I was honestly fearful of ownership. Ownership would be an anchor and would take away my precious freedom. I needed to find out what was in the deal for me.

It is also fair to say that the Cousins had reservations of their own. They met for dinner after that famous shareholders' meeting and came to agreement around one issue: that they needed to hire a lawyer to find out what was going on with their inheritance. The strongest suggestion as to whom to hire came from Charles's daughter, a college professor who possessed a very strong personality. She proposed the attorney who had handled her divorce settlement . . . and in the end the Cousins agreed with that proposal. The attorney she proposed was Guy Hild, a partner with Katz, Teller, Brandt & Hild. It was a great outcome for the employees of Messer. First, it moved the conversation from personal hurt to professional purpose, lowering the emotions on all sides. Second, Mr. Hild had a clear understanding of the options for the future and kept his focus upon achieving a positive outcome for all involved; he was a good lawyer. Third, Mr. Hild was assisted by an associate of the firm who was both smart and effective: Mark Jahnke, who is now Managing Partner of Katz, Teller, Brandt & Hild and a good friend of Messer.

In due course there was a meeting between Mr. Hild and Messer's lead counsel, Mr. Leonard Weakley. You may recall that Mr. Weakley was that young attorney who went to Washington with Mr. Messer in 1932. By 1988, Len was a senior member of the law firm Taft Stettinius & Hollister and an esteemed member of the local bar. Len advised Al to take the matter very seriously, which resulted in my being invited to a meeting to discuss just how credible the employee ownership option might be.

Now things were heating up for me. I had been enthusiastic in presenting at the shareholders' meeting, but was I truly committed to make this work? This is the point when I convened the partners who would guide me every step of the way and who remain my closest advisors today. I called a meeting with Tom Keckeis, Kathy Daly, and Bernie Suer.

At that first meeting we talked about our goals. No one brought up wealth, glory, or titles. We weren't built that way. We all were workers at heart who enjoyed our work. We had no idea whether there was a financial model that would allow us to buy Messer; but we did have a clear idea of some things we would fix if we ever got to be in charge. I left that first meeting with a list, not to share with the Cousins' attorneys, but as clear motivation for having some sort of conversation about the future.

For us the three motivating factors were:

1. Information. We wanted to have information about the company's plans, progress, and results that would allow us to make informed decisions about our own futures.
2. Logical connections. We wanted to understand the cause and effect between inputs (the preparation, energy, intellect, and commitment that we brought to our jobs) and outputs (the recognition, promotions, and compensation we received).
3. Equity. It seemed to us that accidents of birth, accidents of timing, accidents of location, and acts of creative explanation trumped good old-fashion effort and intellect.

These were not new concepts, nor were they set out in a neat list. Rather, they were the sum of years of frustration, most of them normal for any employee. We actually felt great pride in Messer and, for the most part, trusted and respected our bosses, but we were not in the habit of talking about the company in a positive way. Because we were not informed as to the plans for the future, we were constantly frustrated by trying to fill in the gaps. When we were unable to get real information, we would clump up (meaning gather, usually at some bar) and speculate. It is the nature of groups of human beings to speculate more about negative possibilities than to focus upon positive outcomes; better to be armed for the worst and pleasantly surprised than the reverse. After we had traded stories two or three times, some poor person would begin to believe that the speculation was true and would unload on the boss—which always shocked the boss and generally had a negative outcome for the employee.

Armed with a list of frustrations and an attitude that we employees should have a voice in whatever happened, I went off to my first meeting with the Cousins' attorneys. That first meeting was spent getting to know each other and discussing options. Attending were Mr. Hild and Mr. Jahnke (representing the Cousins), Al Berndsen, and a lawyer named Tom Heekin. I came with my attitude firmly in place; and Al came with his attitude that since he was the trustee, he would do whatever he damn well believed was in the best interest of the company . . . especially if it made somebody mad. The saving grace was that the lawyers on both sides came to the room as professionals seeking a solution; and the special blessing was that Tom Heekin was the person who was there to represent Messer's interests. Mr. Weekly had evidently concluded that, given his age and tenure, managing two construction guys with attitudes was just not his cup of tea, so he bestowed that dubious honor upon Mr. Heekin, some twenty years his junior and prepared by his service as a judge to deal with personalities. It is fair to say that Tom's quiet, effective demeanor, quick understanding of facts, and ability to suggest solutions was the glue that held things together over the following stormy months.

Over the next few months, Al and I came to meetings ready to defend against outcomes we didn't control, while the calm people in the room tried to listen and suggest alternatives. Because of the powerful advantages provided for ESOPs by the tax code, it quickly became obvious that employee ownership would be a preferred option. What was not obvious was what the value of the company might be, and even less obvious was the ability of the employees to actually purchase at any price.

A SHORT HISTORY OF ESOPS

The particular form of employee ownership known as an ESOP, where shares of company stock are purchased and held by a qualified retirement trust for the benefit of employees, is the product of some creative thinking by a social economist (my descriptor, not necessarily his) named Louis Kelso. Mr. Kelso, through his work as a lawyer, helped with ownership transitions of some companies and came to believe that the only truly sustainable form of capitalism would be capitalism at the personal level—allowing workers to earn access to capital through their work. Mr. Kelso felt that access to capital would be transformational for workers because hourly effort is not scalable (you have to be there to perform), but capital is scalable over time and might grow based upon great ideas and economic drivers in addition to effort hours. Mr. Kelso was convinced that Congress should act to support employee ownership as a strategy for maintaining the American way of life, and he found in Senator Russell Long a ready believer who was powerful enough to get legislation passed in support of these new ideas. Beginning in the late sixties, Senator Long pushed through Congress a number of pieces of legislation designed to incentivize business owners to sell part of their companies to employees and designed to support the success of those employee-owned enterprises. ESOPs were formed for all kinds of reasons, from the altruistic ("I just want to do something for the employees who have worked so hard") to the very selfish ("The company is dying anyway; the ESOP lets me get my money off the table and leaves the shortage to be divided among the peasants"). Some ESOPs flourished and some flamed out. At this writing, there are more than ten thousand ESOPs in the United States, representing millions of employee owners.

In 1988, when the discussion of the Messer ESOP began, the primary tax support for ESOPs included:

- **Shares sold to the ESOP could be reinvested in any qualifying equity, deferring capital gains tax until that second equity was sold. This right to "roll over" one's ownership interest in a company, where the initial investment was low and long ago and therefore the "tax basis" was very low, had the potential of creating 15 to 20 percent more return for the sellers.**

- **Banks were given a strong tax incentive to lend to ESOPs, which were mostly led by employees who did not have a strong capital base. Banks that lent to an ESOP received a tax credit of 50 percent of their taxable earnings from the transaction, making loans to ESOPs very profitable.**

- **For debt that resided in an ESOP, both principle and interest could be repaid and deducted as a business expense, reducing the company's taxable income.**

Of course, there were lots of rules around who could sell, who could buy, what price could be paid, and how the repayment would need to be structured. However, the bottom line was if you could get a good ESOP designed, and if the employees supported the process, both the buyers and the sellers in the transaction would benefit.

Once the opportunity of the ESOP was understood and the Cousins' lawyers had concluded that the other options were not likely to be as beneficial to their clients' interests, the focus of the meetings turned to value. The interesting challenge here was that previous valuations of Messer had been performed for the purpose of estate taxes, and while all such valuations are supposed to follow the guidance set forth in Revenue Ruling 59-60 (I am amazed at how much I came to know about the tax code during this process; I felt that I should have been awarded some sort of "Cub Lawyer" designation by the time of the closing.), the fact was that all previous efforts with regard to the value of Messer shares had as a goal supporting a low valuation in order to minimize estate taxes. Clearly, the sellers desired a new approach, and, quite honestly, the buyers didn't have a clue. In order to create some sort of progress—and maybe to burn off some energy—we embarked upon a process to capture the "fair market value" of Messer.

At last I had an assignment that I could do with confidence. I may not have known anything about valuation processes, but I had a lot of practice at the Request for Qualifications and Request for Proposals process. I gathered the names of all of the country's leading practitioners; crafted questions as to experience, expertise, and resources; received hundreds of pounds of brochures; qualified a short list; invited proposals; interviewed the proposers; and came to a recommendation. I am not sure that all of this effort was warranted, but it gave me something to do that felt productive, and it allowed time for everyone to calm down. In the end I recommended, and the group agreed to engage, the valuation practice of Price Waterhouse located in Chicago. They had a partner with deep experience in ESOP valuations, and their low-key Midwestern personalities seemed like a good match.

And then the fun began. We shared financial statements, our strategic plans, and our vision for the future with P-W; and after some appropriate discussion and work, they shared a preliminary valuation draft. When I read that draft I almost lost hope; I didn't know much but I was certain there was no way the employees could gather up enough money to buy at that value. (In the eighties it became normal to project that a company could double cash flows as a rationalization for a leveraged buyout, but that approach just couldn't work for us. We didn't have the influence or credibility with the banks to get the loan, and we didn't have the confidence in our own abilities to take the risk. As it turned out, those assumptions wouldn't work in the eighties either—even though the transaction leaders had reckless confidence and could get the loans—so many of the leveraged buyouts failed.) In an effort to salvage our opportunity, I wrote a letter to P-W captioned, "Thirty-two things you didn't think about in coming to this valuation." The first item on my list was a 100 percent turnover in leadership, because we could not lead a company that was so highly leveraged. Most of the rest of the items were intended to take the air out of whatever optimistic assumptions were within the strategic plan.

My letter worked. P-W reworked their analysis based upon my sobering input and arrived at a much more rational number, at least for me. On the other hand, when the Cousins heard about that number they demanded a meeting. At that meeting the central message to Price Waterhouse was that if this clearly irresponsible valuation ever got published, certain Price Waterhouse partners would learn the true meaning of the word mayhem. (I'll leave you to look it up for yourself.)

Price Waterhouse got the message. The partner in charge requested a meeting at which he laid a check in the amount we had given Price Waterhouse for a retainer on the conference table. He then said, "We are returning your retainer. We were under the mistaken impression that you wished to hire a valuation practice. You do not need a valuation practice; you need a mediation service. If you ever wish to have your enterprise professionally valued, please contact us."

This is one more example of how customer service can create loyalty that is very sticky. When we finally did reach agreement and needed a valuation I would never have considered calling anyone else. Even though he is now with a different firm, we still engage the same professional for our valuation. This may be the single most selfless act I have ever experienced from a consultant.

This invites two comments on consultants in general. First, expertise does not require distance. As you would expect, as soon as I sent out the RFQ to the valuation firms, I started getting sales calls from every sort of business consultant. Companies that specialized in helping with business plans, that specialized in designing deals, that could get us financing, that could handle the negotiations. Most of these firms seemed to have deep resources and impressive resumes, and all of them were in agreement about one main point— that we couldn't do it without them. The interesting fact about most of them is that "it" had some specific meaning to them, which was mostly based upon their experience or attitude. They almost never showed an interest in letting us define "it" because they assumed that they just knew more than us about what "it" should be. It turns out, at least for us, that defining the "it" that we wanted to accomplish was what was most important. We did just fine with local advisors, who had plenty of expertise and, more important, who were willing to let us define "it."

The second comment is about what we learned regarding our relationship with our advisors. It became obvious that the experts could answer any question that we asked; however, in several instances, we realized later in the process that we either hadn't asked the right question or that there were whole groups of questions that we didn't know to ask. This experience informs how we interact with our professional advisors today. We do ask them questions, but we also take time to inform them of our plans and propose that they not wait until we think of a question to ask, but rather that they pose questions (and answers) to us that will help us move toward our goals. The willingness to be proactive with ideas and insight is the difference between a consultant and a strategic business partner.

After almost a year of talking and some real effort, we were back where we started. The two big questions on the table were: "How much is this thing worth?" and "Do the employees of Messer actually have the ability and the resources to purchase the company?"

The conversations were about to move from the theoretical to the very practical.

The first four partners. Bernie Suer, Kathy Daly, Tom Keckeis, and Pete Strange.

GETTING TO A CLOSING
[1989–1990]

The Players
PETER S. STRANGE
KATHLEEN C. DALY
THOMAS M. KECKEIS
BERNARD P. SUER

The year was certainly not wasted. We had learned a lot about ESOPs and a lot about what would drive valuation of the enterprise. The question of what we could pay couldn't be answered by conversations with leaders in the company. There was consistent frustration around the key issues we had identified, plus consistent enthusiasm for a world where the inmates had control of the asylum, but also a consistent lack of knowledge and insight as to where the resources would come from to actually pay for the deal.

We now needed to get to work on some new or differently framed questions. Instead of "What is the fair market value of the enterprise?" we needed to focus on "What is it worth to us?" Instead of "Do the employees have the ability and the resources to buy the company?" we needed to focus on "What can we bring to the game that creates value?" I decided to take the first question first, and to try to figure out what level of value could make sense to us as buyers. I had the advantage of having studied the first sixty years of Messer's performance, so I had some notion as to the relationship between company capacity, focus and strategy, and financial results. I just figured that I needed a model. Since there were lots of companies sold and bought every day, I thought it would just be a matter of talking to an expert, entering our particular inputs, and analyzing the outputs.

I took my problem to Messer's audit partner and advisor, who was a local leader in one of the then "Big Eight" accounting firms, and he agreed to do some pro bono work on my behalf. He carved out an hour on Sunday morning to help me make some projections. He had a nifty computerized spreadsheet. What he did not have was a grasp of the difference between making widgets—where capital infrastructure costs, input costs, labor costs, and pricing theory drive the model—and the realities of how predictable cash flows might be created in the fragmented, volatile, transactional construction marketplace. After my hour was up I thanked him as he went off to tailgate before the Bengals' game, and I went back into my funk.

VALUING NON-OPERATIONS LEADERSHIP

It is time to tell an important story. I said at the beginning of my writing that stories are reframed by the teller. I am not sure that Kathy Daly will agree with the details of this story, but I am certain that she will attest to its essence. Kathy and I met in 1970 at the Procter & Gamble Miami Valley Laboratories Project where she was a newly hired project accountant (the first woman hired to work at a project in the history of Messer), and I was in my third year of engineering school working as a co-op doing field layout. I was younger by a few years and dumber by an order of magnitude, but I had the advantage of being an old-timer at this construction game. I grew up in construction, had me a green pickup truck, and I knew damn near everything. She showed up the first day in a dress suit and high heels, because that's the way she had dressed when she worked as an accountant at Shell Oil, and she drove to the site in her stylish little, yellow convertible. I wasted no time in setting her straight. I explained that there are two basic groups of people in the world: producers who create money—that would be operations people like me, and bean counters who count the money that the first group makes—that would be accountants like her.

We didn't get along very well—for many years.

Fast forward to that Sunday morning when I left the Big Eight partner's office. I went down to the lobby of that high-rise and found a pay phone (these were the ancient days before cell phones), and I called Kathy at home.

"Kathy, this is Pete."

K: "What do you want?"

P: "Remember that conversation we had the first day you were at Messer?"

K: "I can't believe that you are already drunk at 10 a.m. on a Sunday!"

P: "Kathy, you need to get this picture into your mind. I am on a pay phone and the cord is just long enough for me to get down on my knees and say: I really, really need the help of an accountant!"

This was my first step toward recognizing the value of diverse talents and skills. Over the next year we spent most of the weekends together working on our model. Each time we tried a new combination of assumptions and projections, we called it a "Run." With Run 54, we found a plan that we could use to buy a company.

While Kathy and I were busy developing a cash flow model that reflected the realities of construction, we also started addressing what to do about the non-construction assets. These assets fell into three groups:

- Real estate that Charles had developed or purchased when his focus was on protecting the family fortune. Al Berndsen had succeeded in refocusing the company on construction, but he had not been able to make clean sales of all of the real estate. Two properties, a garden style apartment on the west side of Cincinnati and a high-rise apartment near downtown, were partially paid for by the purchasers, with substantial balances under land contracts still due over the next few years.
- The one thousand acres of land in Clermont County that Messer had extracted from the failed partnership with the brothers, which carried a mortgage of $3.8 million requiring regular, substantial loan payments.
- The second mortgage on The Beach Waterpark, which was carried on the balance sheet as a $1.5 million asset but which produced no cash flow.

Obviously, as employees of the construction operation and as individuals who would be hard-pressed to borrow capital, we had no real interest in purchasing anything not directly related to construction.

In one of the single worst ideas I had ever put forward, the company had tried to sell the Clermont County land at auction a few years earlier. We engaged a national firm; advertised in national and international papers; and we talked ourselves into believing that we would succeed. On the night of the auction, we were in a ballroom at a downtown hotel. We had a team of auctioneers; we had lawyers ready to bind the deal; we had a band to celebrate the sales; we had everything except buyers, as only a few farmers showed up to bid. Al Berndsen stomped out after the first parcel brought only a bid from a young farmer who wanted to own some land before he proposed to his true love. It was not a good night, and it convinced us that Messer would be a major farmer—we had a tenant farmer who rotated corn and soybeans on the 888 tillable acres—for many years to come. We had no new ideas about how to divest of this asset.

We approached the buyers of the two apartment properties proposing that they accelerate payment. They had no interest in modifying their contracts.

So that left The Beach. I called the owner of The Beach and informed him that we intended to sell our second mortgage, even at a discount, and that we would prefer to sell it to him. That's when I got one of the real shocks of the year. He asked whether I was in my office, then suggested that I hang up the phone and wait for him to appear in person. When he arrived, he informed me that he had been putting off the uncomfortable call to inform Messer that The Beach could no longer service the debt on the second mortgage.

This series of conversations was more than just discouraging. We didn't know much about borrowing capital or how we would gain surety support; but we knew that non-performing, non-operational assets would not be an advantage.

LEADERSHIP COMMITMENT

Our conversations with the attorneys for the Cousins had been ongoing because they were by this time vitally interested in a credible offer and an early closing. I took the matter of the non-performing assets to the bargaining table and suggested that we employees could purchase the operating assets, but we would need to deed the non-operational assets over to the sellers. This concept became part of an ongoing wrangle as both sides pushed for the other to solve the problem. At one point, the conversations became so heated that the Cousins' attorneys suggested that they would like to talk with other employee leaders—obviously hoping to find someone smarter and more reasonable.

─────THE INDIANAPOLIS STAR─────

SUNDAY, NOVEMBER 8, 1987

AUCTION
1,000 ± ACRES
Cincinnati (CLERMONT COUNTY) Ohio
to be held
7:00 P.M., WEDNESDAY • NOV. 18
PREVIEW: 1:00 to 6:00 P.M. TUESDAY • NOVEMBER 17
AT PROPERTY LOCATION — OR DRIVE BY
at the
DOWNTOWN WESTIN HOTEL • FOUNTAIN SQUARE
PROPERTY LOCATED 20 MINUTES EAST OF CINCINNATI
LOCATED NORTH OF THE FORD TRANSMISSION PLANT
OVER 3200 FT. OF FRONTAGE ON S.R. 276 — NEAR EASTGATE
MALL ON S.R. 32

PROPERTY TO BE OFFERED IN TRACTS RANGING
FROM 12 ACRES TO OVER 100 ACRES IN SIZE
BUY ALL OR BUY PART

DIRECTIONS: I-275 NORTH TO EXIT 63B BATAVIA EXIT EAST
PAST EASTGATE MALL AND FORD TRANSMISSION PLANT
TO HALF ACRE ROAD EXIT NORTH TO S.R. 276, TURN LEFT
TO THE PROPERTY ON RIGHT BETWEEN S.R. 276 HAWLEY
ROAD, JACKSON PIKE AND SHARPS CUT OFF.
TERMS: A 10% CASH OR CASHIER'S CHECK DEPOSIT RE-
QUIRED DAY OF THE AUCTION. CLOSING ON OR BEFORE 60
DAYS, 10% BUYERS PREMIUM.
CALL FOR COLOR BROCHURE

America's #1 Auction Team
Auction Company of America
Miami • Denver • Chicago • Dallas
Auctioneers • Realtors
Jim Gall, Founder and Champion Auctioneer
1-800-643-0808 or 305-577-3322
In Cooperation With Bill Holland Auctioneer and Sanders Home Corp.

**Clermont County land auction advertisement. These ran
nationally in the *Wall Street Journal* and internationally in *The Economist*.**

93

The timing was perfect for me to have the right answer. The request for more reasonable employee leaders came on a Monday afternoon. Over the past weekend, Bernie Suer had broken his leg at a family picnic. Bernie's approach to dealing with this accident was to have his wife deliver him to the office at six o'clock that Monday morning, so no one would see him crawling up the stairs to his second-floor office. I shared this story with the lawyers and suggested that Bernie was representative of the calmer, less-obsessive option. They dropped the subject.

We kept meeting every week but everyone was frustrated. At one point I ran out of new ideas and simply suggested that I earned my wages by the day and that since this didn't seem to be working out, I would be about the business of finding new employment. Now, one of the things that convinced the Cousins that someone was stealing from them was the fact that non-family members had contracts. They knew about Al's, Loren's, and others' contracts, and when they started the conversations after the shareholders' meeting they were sure that Al was illegally giving away their fortune by giving contracts to people like me. That was not the case as none of the younger generation had contracts so the matter was dropped.

However, on the evening after I made my threat to quit, Al received a call from one of the Cousins to inform him that it was their conviction that he had acted irresponsibly in not having me bound to a contract and that if I actually quit and the deal fell through, they intended to sue him. As I said before, we were very lucky that the Cousins hired good lawyers rather than representing themselves.

WE'LL WRITE IF WE FIND WORK

Each meeting with the Cousins' attorneys required a debrief with the leadership team. We would meet, usually at Bernie's house or Kathy's condo because they were closest, and over much bourbon I would share what was going on and receive lots of questions and council. Since Tom, Bernie, and Kathy were now doing most of my work as well as all of their own they had a very strong identity of interests around getting this deal done.

When we hit this rough patch we began to question whether doing a deal with the Messer family was really possible. In one meeting, we talked about starting a company. Very quickly Tom, Bernie, and I framed up how that would work, who the customers might be, and

the names of the key builders we would need. Then Kathy asked the question, "What about me?" I don't recall exactly who gave the answer, which was, "We couldn't afford someone like you at the start, but we will let you know when to come over." I do clearly remember Kathy's loud and angry response to that statement. "So I am to be left behind at Messer like baggage to be claimed at your convenience? No way!" We went back to work to get the deal done.

With the model in place that showed how the construction operation could create positive cash flow, we set about the business of finding the offering price. We did not have agreement on who would own the non-construction assets, but we needed to have some sort of target so we could start to figure out where the money was going to come from.

By this time it was clear that an earn out—where we would pay the Cousins off over time—was not an option; the Cousins just didn't have enough of an identity of interests to stay connected over a long period of time, and we employee leaders could not conceive of their continued involvement in decision making for the company. If we were going to move forward, we needed to have a single closing and buy the company. For many reasons—some practical and informed, some emotional and unsupportable—we focused upon the book value of the company, about $7 million, as the target price.

In looking for sources of capital we assumed that we would be able to use the money in the profit sharing plan as part of the purchase funding, since the plan was originally designed to allow conversion to an ESOP. We projected that the profit sharing plan would have assets of at least $1.25 million by the end of 1989. We then turned to the question of what the employees could bring to the table. We targeted a group of twenty senior managers, for the simple reason that we did not want to pick and choose among individuals, and Securities and Exchange Commission regulations would require an expensive securities registration if we had more than thirty-two investors. We went down the list of job titles starting at officer and stopped at project manager (now senior project executive) because the number of people with the next title (superintendent) would have taken us well over the limit of thirty-two. We set a target of $1 million for this group. We felt that we would need to show a substantial commitment to the sureties if they were to continue providing surety credit, and the average of $50,000 per investor seemed doable, especially if we could use our individual retirement accounts (IRAs). That left us with $4.75 million to find—and to pay back after we found it.

THE BENEFITS OF LOYALTY

One of my firm assumptions was that we would not be able to borrow that sort of money from Star Bank—our lead bank since that event involving the carriage house project more than seventy years before. During a meeting with George Elliot, the head of the bank division that managed our account, I explained that this was just a courtesy conversation since I knew they would have no interest in lending money to a bunch of employees who wanted to buy a construction company, and that we anticipated that we would need to go to Chicago or New York to find a loan.

I managed to insult George to his core. He told me in no uncertain terms that First National was as smart, as committed, and as creative as any of those big city banks. This was one more stroke of luck. By accident, my abrasive personality motivated George and others at First National, not just to loan us the money but to stretch to create a new model, allowing us to guarantee only a portion of the principle so that we could present a stronger balance sheet to our surety in support of bonding. This is one of many examples over Messer's history when our being loyal to a business partner over many years paid a big dividend.

With these pieces in place we went back to the model to figure out what we could buy with $7 million. We didn't want the non-construction assets, but neither did the Cousins. We finally came to a deal that included the employees purchasing the Clermont County land and the two apartment building notes, while ownership of The Beach note would be transferred to the Cousins prior to the closing. On November 7, 1989, we reached agreement on the purchase of Messer by the employees and set a targeted closing in three months. Now all we had to do was to turn the conceptual plan into reality.

One of our first challenges was to create a structure that would assure the bank that they would receive the portion of the cash flows required to repay the debt. The solution was to create two classes of stock—preferred shares, which would be held in escrow by the bank and which would receive a dividend to be used to repay the loan; and common shares, which would receive no dividend and would be owned by the ESOP and the employee investors. Fundamental to this process was reaching a fair allocation of equity ownership that would make the two classes of shares "equal" since both classes would be held in the ESOP, and the ESOP, as a qualified retirement plan, must pay fair value for both classes.

The result of our analysis was that the cash investors—the employee investors and the profit sharing plan—would receive 54 percent of the company equity for their investment of $2.25 million, and that the investment funded by debt would receive 46 percent of company equity plus a stream of preferential cash flows through dividends. I was very proud of this math, which was done with the guidance of some experts from Houlihan Lokey Howard & Zukin in Chicago. I was proud until an associate of the same firm of experts, but from the Los Angeles office, called me at 6:30 on the evening before Thanksgiving 1989. This confident and brusque fellow informed me that he had looked at our transaction and it just wouldn't work. The asset allocation was indefensible and would not be approved by the Labor Department. We talked (argued) for a couple of hours. I hung up, looked at Kathy (who had come in to see what all the yelling was about) and said, "Evidently our only choice at this moment—since the deal is now dead—is to go get good and drunk." This conversation caused a late arrival home and a lousy weekend, but it did not kill the deal.

It turned out that we really did have the math right. This led to one of my favorite things to say to financial people and accountants: "This ain't math. Math is finite element analysis and Lorentz transformations. This is just arithmetic: adding, subtracting, multiplying, and dividing. No matter how many rows and columns our spreadsheet contains, we ought to be able to get this right."

THE POWER OF TRUST

As I have said, the Messer profit sharing plan that was established in 1984 included language that allowed the trustees to vote to transform the plan into an ESOP and to purchase company stock with the assets. Since the trustees in 1989 were Al Berndsen, Jim Perin, and me, my assumption in putting together the transaction was that this money was a given. Another piece of good luck and good counsel came when Tom Heekin, that calm attorney who had taken over as Messer's lead counsel, challenged this assumption.

As far as I can remember, in our entire relationship now spanning more than twenty five years, Tom has never raised his voice even to a level of insistence, much less anger. He just keeps talking about something he calls his stomach test until you begin to understand that you are misguided and need to mend your ways. In this case he did not try to convince me that we couldn't proceed without taking our plan to the profit sharing participants. Instead he started asking me questions about what my priorities were for employee ownership, which I answered with the usual line about deeper employee engagement based upon better communication and informed support. To which he posed the question, "If you are committed to deeper communication and informed support, why aren't you sharing the plan with the employees?"

That question made sense to us so we called a meeting of the profit sharing plan participants. The Cousins did not want this meeting to occur because they saw it as my sharing their personal financial information. They warned me that if we held the meeting and the closing did not occur, they would hold me personally responsible— which was just the stimulus I needed to become enthusiastic about the meeting. We held the meeting at a local restaurant and had nearly 100 percent of the ninety-nine profit sharing participants in attendance. Each attendee received a packet that contained the goals for employee ownership, the financial projections for the company after the purchase, and a form that could be used to direct the trustees to use an individual's profit sharing account to purchase company shares.

One of my proudest moments was when we counted the results a few days later. Of the ninety-nine participants, ninety-eight had signed the form requesting that 100 percent of their account be used to buy company shares (the last participant wrote a note saying that it was probably a good idea but he just couldn't understand the math). Did the other ninety-eight really understand convertible preferred shares and agree with my presentation on the details of the plan? I doubt it. However, I am certain that they understood that they were being trusted with information at a new level.

That was the moment when I understood that trust is to human performance what water is to human health. If employees get a lot of trust every day, they might thrive. If there is no trust, it really doesn't matter how smart the plan is on paper, you do not have a sustainable enterprise. Over the past twenty years, we have shared everything we know about our plans and resources, and I cannot cite one example where the sharing has had a negative impact. People work better when they are trusted and when they can make informed decisions about their own futures.

Our next set of conversations was with our surety company, the United States Fidelity and Guarantee Corporation (USF&G). Here we had the advantage of my having been in home-office meetings with their senior leaders, plus we had the good fortune of working with one of their most able and influential regional managers, Andy Everett. The combination of Al's personal influence, Andy's support and coaching, our improving performance and unique risk analysis report, and, not least important, the fact that twenty senior managers were putting $1 million of their own money into the deal led to a continued commitment of $100 million in surety credit after the closing. This was critical if we were to continue to grow the company.

THE COMMITMENT THAT CHANGED THE WORLD

The twenty employee investors were mostly young, and all were new to the concept of buying companies. Twenty people would be putting their own money in the deal to make this happen. The amounts ranged from $10,000 to more than $200,000; however, every commitment was significant for the employee who made the pledge. Many took second mortgages, deferred major purchases, or committed their IRA retirement funds in addition to their profit sharing.

To meet the requirements of the bank and the bonding company, the rules of the game were tough. The employee investors had to commit for ten years without any right to withdraw. And, in our enthusiasm for raging democracy, we added another even more demanding rule: we asked each employee investor to sign a restrictive stock agreement that gave the ESOP trustees both the right to vote his or her shares and the right to repurchase those shares at will. The employee investors put their funds at risk just to support employee ownership without many of the normal rights of ownership.

This commitment was singular in its level of support for the concept of widely shared ownership; and, more important, this commitment laid the foundation for Messer's differentiating advantage in attracting talent. Unlike many companies (even those with ESOPs), Messer doesn't have to explain to recruits that special deals exist for some owners because of their timing or tenure. Everybody at Messer is in the same box: same expectations, same accountability, and same opportunity. If someone in the 2013 recruiting class has the aptitude, energy, and commitment to become CEO, he or she

should retire from Messer with the same proportional ownership as Pete Strange or Tom Keckeis. This open marketplace for talent sets Messer apart; and the fact that the employee investors were willing to sell their shares created the opportunity for the ESOP to take advantage of the Subchapter S tax election—allowing the deferral of federal income tax until an ESOP participant retires, which adds significantly to the value of ownership. When you see an employee investor, thank him or her.

As we moved toward closing, it became obvious that we would need to prepare literally hundreds of documents to get the deal done. Over sixty years, many details had developed around the Messer companies that would have to be dealt with, and many representations needed to be made to lenders, sellers, and others. We needed to have a corporate secretary in place, and Jim Perin had retired earlier that year. Kathy was the obvious choice, and I was able to get Al's agreement to name her secretary of the corporations.

There were, however, many concerns to be dealt with regarding the potential that the closing might not occur, and we needed to manage the Cousins' sensitivities to our actions. After sorting all of that out, in the end I was able to go to Kathy with the good news that she was being elected corporate secretary but with the following ground rules: she could have no raise, there would be no announcement, she was not allowed to tell anyone, and she needed to check with me before she signed anything that might be seen by the family. She was thrilled by the intention but not too impressed by the execution of the honor. However, her only real sticking point was the title secretary. She told me that she had been defending against that title for her entire career and that she expected the title vice president at the closing.

On January 17, 1990, the Messer ESOP was formed and immediately purchased 92 percent of the shares of Messer stock from the Messer family trust. The employees received the right to own their future, and the Cousins received their equity, with the right to defer capital gains on the sale indefinitely if they chose. Plus, the Cousins received as a dividend the second mortgage on The Beach Waterpark, which they were able to sell a few months later. About thirty days later, a second closing was held with the non-family owners for the remaining 8 percent of the shares. In the end, the employees purchased the company for about book value, which is consistent with the general valuation of contracting firms. Over the following five years, the installment sales of the apartment buildings were completed, and the Clermont County land was sold for a small gain.

Employee Investors

- Pete Strange
- Ed Miller
- Tom Keckeis
- Bernie Suer
- Kathy Daly
- Allen Begley
- Don Beiting
- Verl Campbell
- Larry Keckeis
- Neil Fry
- Bill Krausen
- Jim Hess
- Mark Luegering
- Dave Miller
- Craig Reese
- Bill Rutz
- Rick Zoller
- Bob Wassler
- Steve Eder
- Frank Messer

The employee investors. If you are an employee owner
of Messer, and you see one of these people, thank him or her.

NON-CONSTRUCTION ASSETS

Surety companies understandably are negative about construction companies investing in non-liquid assets that cannot be brought to bear on project needs. For many years, a central part of the conversation between Messer and USF&G was the value and the exit strategy for our non-construction assets. When the sale of the Clermont County land was completed, we purchased fifty small glass containers, filled them with a handful of dirt from the property, and had engraved on the sealed lid: "This box contains 2 percent of the undeveloped land owned by Messer as of August 15, 1999." We gave one of these boxes to each of our credit grantors. That statement is, of course, no longer true because we have the resources and expertise to both invest long term and support our construction operation, but at the time, the sale of the land was a real milestone.

The Messer plaque. The only way to own this object is to be a senior manager/partner in the greatest construction company in the world.

The employees' purchase of Messer received good coverage in the press. We were justly proud and wanted to share our story. One day I received a call from a very nice person representing the *Cleveland Plain Dealer*. He seemed to share my enthusiasm for employee ownership, and I was happy to talk with him. As we completed our conversation he mentioned that he knew we had a lot of non-family members to deal with, including members of Senator Robert Taft's family. Senator Taft, a good friend of Frank Messer, had invested in the company early in its history, and when his shares were repurchased he retained 125 shares as a keepsake. I happily recounted the story of having to hunt down a dozen relatives, each of whom owned less than ten shares of Messer stock. The reporter casually mentioned that Bob Taft must have been easy to find because he was a Hamilton County Commissioner. I laughed and agreed. The next day an article ran on the front page of the *Cleveland Plain Dealer* under the headline, "County Commissioner Owned Stake in County Contractor." For a long time I wouldn't talk to anyone from the media; and even today I believe that it is best to think twice before you speak once when dealing with the press.

M

The Messer partners in 1992.

7

STEERAGEWAY
[1990–2012]

The Players
MESSER EMPLOYEE OWNERS

On January 17, 1990, after the closing with the Cousins, the employee investors and many others who had helped with the purchase transaction had a big
party. It was a great party—at least the part I remember was great—but after
the party we were left with the question of what to do next. (It was around
this time that Kathy came into my office and asked the question we asked of
each other many times: "When do you think the grown-ups will be coming
back?") In many ways, our focus had been on the deal, not on leading and
growing the company after the deal was done. We needed a plan that all of
the employees could understand and support. On February 11, the twenty
employee investors came together at Camp Joy to focus on, "Where do we go
from here?" In May of that year, we gathered at the Marriott in Lexington to
finalize and adopt our Mission and the core of our strategic plan. A lot of us
had new jobs, and all of us had a new, shared responsibility—to make Messer
successful for the employees who had trusted us.

There was a lot going on and most of it was positive, but some mistakes I
had made were having a real impact. My first and biggest mistake was that in
my enthusiasm for doing the deal I had communicated that every participant
would realize his or her goals if we could just get control of the company.
That was not true and in hindsight I should never have acted like it was
true. The harsh reality was that the individual employee investors had a wide

range of talents and experiences, and they had, with my support, some really lofty personal goals. Many, if not most, of them believed that I had promised them promotions and raises if we could just get this deal done.

The reality was that we had committed to some strict fiscal responsibility and that there were a lot of lenders and surety people watching to make sure I did not reward my friends at the expense of the company. When it came time to make the organization chart for leadership, it was lean, and some folks rightly felt left out. We lost some of those good folks. Losing those leaders as a result of my hyperbole taught me that hope is not a strategy and that honesty and clarity are an absolute requirement for leadership.

GETTING ELECTED

The first organization chart was so lean that there was literally only one person for each area of responsibility, which was interesting because two people—Tom Keckeis and Bernie Suer—had stepped in to lead Operations while I was off negotiating with the Cousins. My excuse for ignoring their shared leadership is that I was focused upon being lean in the eyes of the lenders, and I knew that, because of his friendship and his level of commitment to making employee ownership work, I could ask Bernie to wait without fear that he would up and quit. So I made my organization chart and began sharing it with the individual employee investors. Bernie didn't say much about the omission of his name, but the people that Bernie supported had a lot to say. A large number of people questioned why their boss and advocate was omitted and wondered aloud whether I understood that advocacy is a two-way street. Bernie had stood up for them and supported their growth; they now stood up for Bernie. Before too long, I had a new organization chart: one with two Operations VPs. Once more, proof that loyalty is not a trait; loyalty is the reflection of how leaders treat the people they support. Bernie Suer was not selected to lead by Pete Strange—Bernie was elected by those whom he served. Helping others get better results, make better decisions, and think longer term is the short path to leadership.

It was recognition of the power of information and the loyalty that could be developed through informed support that led us to adopt Messer's organizational structure. No matter what the organization chart said about authority, the facts were that the strategic plan needed to be understood and supported by all employees and, even more important, each individual employee wanted and needed an advocate who would "be in the room" when important decisions were made about the future. Largely in response to this reality, but also in part as a reward for the unselfish commitment of the employee investors (and a little bit as a reaction to having worked in the "King and his court" world for twenty years), we decided to organize the company operationally as a partnership. We use the term senior managers and partners interchangeably, not as legal descriptors but as the definition of a key level of responsibility in the company. A Messer partner is "in the room" when strategic decisions are made; his or her voice is heard in coming to those decisions; and he or she has a direct responsibility for making the business plan real for other Messer employees and for supporting the growth of everyone they touch. The Messer senior management group has grown from the twenty employee investors to more than one hundred leaders from every part of the company. We have leaders in the partnership group who graduated first in their class from engineering school and have an advanced degree; and we have leaders who came into Messer through the craft tradition and have found their education through very non-traditional paths. What every one of those leaders shares in common is a willingness and ability to help others make better decisions, get better results, and think longer term.

> **The Messer senior management partnership was our first step toward creating steerageway.**

The year 1990 was good for Messer. We put in place more than $100 million of construction and, partly driven by the favorable resolution of some potential problems on a large project in Columbus, we achieved record profitability. We were on our way; and then the economy went south on us. In 1991 we saw the beginning of a recession, which made selling work more competitive and compressed margins. And, while those external factors were challenging our projections, we had a major project problem develop.

Looking back even now, the problem wasn't really our fault. We were asked to complete a challenging project on a very tight schedule, and we felt that we were really clear in communicating to the owner that the focus on schedule and the many changes that were being initiated were having an adverse impact upon cost. Whatever the disconnect was, at the end of the project when we presented the cost changes, the customer's representative was appalled. He had no expectation that we were off budget, and, more important, he had not prepared his bosses or the board for the need to fund overruns. He understood that he and others had made changes and pushed on schedule; however, he expected Messer to take responsibility for all of the project goals—quality, schedule, and cost.

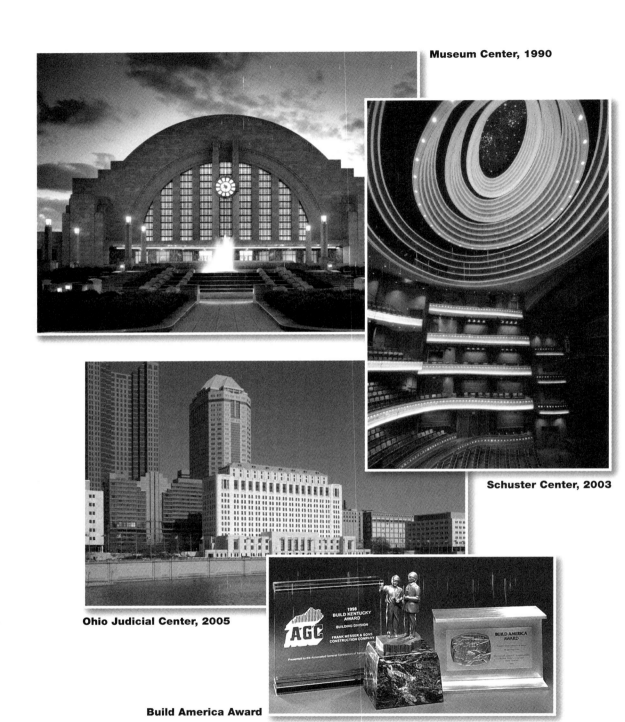

Museum Center, 1990

Schuster Center, 2003

Ohio Judicial Center, 2005

Build America Award

BUILD KENTUCKY, BUILD OHIO, AND BUILD AMERICA AWARDS

As has happened many times since, the leadership of Messer had a decision to make. We could be about the business of proving that we were right, at the expense of losing a customer; or we could do what the customer thought was right, at a severe cost to our bottom line. At the end of the day, we concluded that we would absorb more than $1 million in disputed costs to retain the customer. While it had an impact upon an already tight year, our conclusion was then, and our conviction remains now, that long-term customers are always more important than short-term project results; and our related conclusion is that we must communicate with customers with great clarity, or accept the consequences.

> **Focusing upon long-term, strategic relationships with our customers was the second step toward creating steerageway.**

Long-term thinking was the biggest transformation caused by the ESOP. We now had a promise to deliver on when an employee retired, whether that event occurred ten years, twenty years, or thirty years out. Because we as new leaders did not have contracts that focused us on getting our share of this year's profits, we could focus on investing to protect our ability to deliver on that promise. It can be frustrating to settle a dispute when you feel strongly that you are right; but investing in the future has proven to be a sound strategy. Long-term thinking causes you to realize that there isn't any big win that is worth mortgaging the future.

Just as important, thinking longer term transformed our relationship with our employees. Access to capital is an elegant opportunity, but it really doesn't work unless you stay in the game long enough to allow the power of capital growth to kick in. No lesser light than Albert Einstein is reputed to have said something like, "compound interest may be mankind's greatest invention." Messer and its employee owners shared a new identity of interests. The employee wanted growth over a long period of time to grow the value of his or her capital; and the company wanted a long-term payback on its investment in developing the employee. In response to these new realities, Messer's leadership began to focus on a "career-length" planning horizon.

- The employee would be empowered to have a voice in changing Messer, rather than needing to change employers. It is this notion that led to thinking of the senior managers as partners: each with access to deep information about company resources and results; each with a voice in shaping future strategies; and each with a responsibility for making those plans real for the employees around them.

- The company could invest in employee development with confidence that the investment would have a long-term value. It is this notion that led to creation of the Professional Development Group and the company-wide support for individual education, growth, and development.
- Most important, this identity of interests between the company and its employees led to a strategic discipline around growth. If career-length relationships served both parties well, then it made no sense to be tactical about hiring. Construction companies have traditionally matched their growth to the market, growing at whatever rate they could sell work, and cutting back through layoffs if the flow of work slowed. Messer decided to take a new path: to focus upon growing capacity through growing leaders, then to deploy that capacity toward project opportunities.

The forty-year, career-length planning horizon and putting long-term capacity building, customer relationships, and resource development ahead of short-term gain became a defining element of Messer's path to steerageway.

FINDING PETE A JOB

So who had time to think through all of this complicated stuff about planning horizons, especially in the midst of a recession and the day-to-day realities of leading an enterprise? It didn't happen by my design. We, like most company leaders, were obsessed with meeting our goals and obligations; however, I found myself in an odd position. I had been focused for two years on buying the company. I had learned a lot about structure and strategy, but I really didn't know much about the job of being CEO. So when I didn't know what else to do, I went back to doing my old job—I had at one time been a crackerjack VP of Operations. I started following Tom, Bernie, and Kathy around to make sure they learned their jobs from a master. Ignoring the fact that the three of them had been running the company while I was negotiating with the Cousins, I would give advice, speak for them, ask questions, and, most important, I would point out the many errors in their thinking and actions. I was having a really good time, but I was completely oblivious to the fact that what I was doing wasn't so pleasant for the three of them.

**University Hospital
Critical Care Pavilion**

**The Surgery and Heart
Center at Grant Medical Center**

Xavier University's Hoff Academic Quad

**Cincinnati Children's Hospital Medical
Center—Location S Research Tower**

BUILD OHIO AWARD WINNERS

I tell this story with the deepest gratitude. I think I grew into a pretty good CEO. That never would have happened without the strength and care of my team. About six months after the closing, Tom, Kathy, and Bernie invited me to a planning meeting offsite. I thought I would spend the day informing, correcting, and redirecting them from my higher level of insight into their jobs. They came straight to the point. They said they loved me and they hoped we would always be friends, but they had no intention of continuing to work with me unless I found something to do other than criticizing them.

Like most turning points in one's personal growth, it didn't feel very good to be told that my attempts to help often had the effect of "taking people to zero." We spent the day working through our job descriptions, with a focus on the boundaries between what I could do to help and where I was just interfering. I did not come away cured. They still correctly say that I gather information through cross-examination and that having me in a meeting is an invitation for me to come up with a new idea that runs the conversation down a rat hole. Nonetheless, I left that offsite with a clear idea of what my job as CEO needed to be.

Here, with the benefit of many years of refinement, is my job description for the CEO of Messer. It will continue to evolve—to remain vital, Messer must remain a work in progress—but this is good description of the core.

- Structure: A company is a machine for turning personal assets, abilities, and energy into value—value for customers, value for our families, and value for our communities. Like every machine in the world, our company machine has an efficiency factor—outputs divided by inputs. Every bit of human intellect, emotion, energy, or creativity that is lost, misplaced, misguided, or misused inside our machine, is lost to value. The CEO's job is to make sure we have the most effective machine possible.

- Communication: This is fundamental. A worker can hold a shovel in his or her hands and can dig ditches; or that worker can be about the business of helping to cure sick children. The physical activity is the same, but it is two completely different worlds to live in in terms of engagement, pride, and emotional rewards. We know that every individual who works for Messer is a good, smart person. The job of the CEO is to put every one of those good, smart people into ownership of the opportunity to look beyond the task and to help create value for their customers, their community, and for their families.

- Change: Many leaders talk about change management and getting ahead of change. It can't be done. The really big changes—earthquakes that disrupt communities, tsunamis that interrupt the supply chain, unrest and wars that disrupt societies, and upheavals that ravage economies—cannot be managed or controlled at the individual company level. It is the job of the CEO to create steerageway, to drive the action so that the enterprise can set direction amid the changes that cannot be controlled.

In the spring of 1991, the senior managers came back together to review progress and assess the plan. We started that second meeting with a challenge to the leaders of Messer. All twenty leaders had participated in creating our new mission just a year earlier. There were twenty blank pieces of flip-chart paper on the wall, and each leader was asked to go to a piece of paper and write the mission of Messer. Most got the gist of the message, but no one could recreate accurately the mission that had seemed so important to us when we were setting new direction. There was, of course, a great deal of spirited discussion as to whether getting the intent was enough, or whether we as leaders needed to actually own the words of the mission. In the end we agreed that we could not expect hundreds of employees to get "it"—whatever "it" was—unless we as leaders absolutely owned "it" and were able to share "it" without research and notes. This second planning meeting started us down the path toward understanding a key responsibility of the partners who lead Messer: the responsibility for making the Mission, the Values, and the Strategies real for every employee.

THE CREW CONCEPT

For those of us who grew up around craft work, there is an absolute about crew leadership: no matter how simple the work and no matter how gifted the foreman, a crew leader cannot meet the needs of more than about six to ten crew members. Human beings are just too complex for one leader to meet those needs—the need for attention, for information, for support, for consultation, for affirmation—for more than ten people. If the work requires more

than ten workers, you have to split the crew and appoint a second leader to assure productive effectiveness.

When we move to management, this lesson is often forgotten. Especially with today's mass communication tools, we tend to think that one e-mail to hundreds of employees, if carefully crafted and accompanied by creative media support, will guide those employees to effectiveness. It will not. Communication is not what we send—the power of our speeches, the cogency of our arguments, the force of our data, or the impact of our media presentations—communication is what is received (and acted upon) by the human beings on the other end of the message. And, those human beings have a human need for engagement, appreciation, attention, information, clarification, support, and affirmation. When it really matters, we need to lead with respect for the crew concept.

This responsibility for owning the plan and for making it real for those across the company moved the planning from being an extracurricular activity—to be enjoyed and driven during the annual retreat—to a core day-to-day responsibility of leadership.

When I joined Messer, I noticed that people who worked in the office were chronically late in arriving at work. The official start time was 8:30 a.m., but the accepted arrival time was from 8:30 a.m. to about 9 a.m. I'd grown up with Grandpa, whose mantra was, "If you can't be on time, be early"—usually followed by a snort in my direction, "because, Petey, it takes you longer to come up to speed than most people." In the field where I grew up, work started at 7:30 a.m. I would arrive at Messer at about 7:30 a.m. It took me a while to realize that everyone else wasn't just lazy or irresponsible; everyone else was acting like the boss. Our boss had embraced arriving late as both a perk of his office and as evidence that he was smarter than the rest of us. He would arrive promptly at 10 a.m., even for a 9 a.m. meeting; and by his action he communicated that this is what smart people do. Interestingly, as slow learners like Tom, Bernie, Kathy, and me—who always arrived early—took over more responsibility, the vast majority of the Messer employees came right along with us, without the need for new policies, announcements, or disciplinary action.

Messer's path to steerageway is based upon the hard reality of leadership. We will get from others about 30 percent of what we demand; we will get from others about 60 percent of what we teach;

but we will get from others 100 percent of what we model. We must decide to live the strategies, values, and mission we want to achieve; then we need to respect the crew concept so that those we touch can truly connect with us and live the model with us.

A good example of the changes that we cannot control occurred in 1993 when the carpenters' union and the laborers' union decided to coordinate their negotiations with Messer to send a message to other employers. Messer had long been one of the region's largest employers of the "Basic Trades"—carpenters, laborers, cement finishers, rodbusters, and operating engineers. Over the years we had focused more on carpenters and laborers for reasons of efficiency; however, Messer was clearly viewed as the employer leader.

For more than a decade prior to 1993, Messer and its unions had been negotiating on cost and subcontracting issues. The rise of large, productive, non-union contractors had put price pressure on union employers. Messer's focus in those negotiations had been on the subcontracting clause in the union contract—the clause that stipulated that Messer could not subcontract to a non-union employer any work that the union trade normally performed. Negotiations had come to a head in 1984 when the unions representing Messer employees had initiated a strike solely against Messer, rather than following the traditional path of striking all of the union employers. This approach, which had been used many times in other industries, allowed most of the union members to remain on the job while a model contract was hammered out with the employer who had been struck.

At that time, Messer was in the middle of two of the largest projects in its history to that date—the Hamilton County Justice Center and the Cincinnati Insurance Office Building—so the work stoppage was timed to have a real cost impact. More important, in 1984, even though Messer had many craft employees who had worked for the company for decades, and many Messer supervisors had come up through the crafts, Messer leadership was just not sure what would happen if we asked the workers to return to work rather than honor the strike. The result of that uncertainty was that after two weeks we went back to the table and signed the master agreement.

We signed that agreement knowing that market realities, not our comfort or our economic goals, would define our future. The reality of the markets convinced us that we could not continue a process where we gave in to union pressure. The result would be that more competitive contractors would take our market share from us. Shortly after the 1984 strike was settled, we initiated communications with our craft employees. What we learned scared us.

At a time when Messer had more than twenty-five active projects, most of our craft employees could name only two or three—the rest simply did not exist as opportunities in their minds. At a time when many of our craft employees had experienced continuous employment for decades, nearly every craft employee believed and feared that he or she would be laid off at the end of the current job—because we only talked about employment in terms of the current project. And, worst of all, almost none of our craft employees had any understanding of how we got new jobs—they believed that selling was done by people in the office and had nothing to do with them.

Obviously, we had some work to do, not just because we feared another strike, but because our craft workers were (and are) the key to our quality and most often are the face of Messer that our customers experience. We needed them in the game with us. So we called a meeting—and almost no one showed up. Going to meetings with bosses after working all day was not a popular idea. We made adjustments and kept working at it until we found ways to reach out to our craft workers.

MESSER POCKETKNIVES

If we scheduled a meeting at a motel meeting room right after work, we were told that the craft workers didn't want to come because they were too dirty. If we scheduled the meeting at 7 p.m. so they could go home and clean up, some of them didn't go home; they went to the bar. Then, by the time 7 p.m. came around we didn't want them at the meeting.

We found the key to this challenge in a symbol of appreciation. For us, the symbol is the Messer pocketknife. On the third or fourth unsuccessful try to have a meeting with all the craft workers, at the end of the meeting I stood up and told the 20 or 30 percent who were there how much we appreciated their time and commitment. Then I stood at the door and handed each one of them a pocketknife engraved with "Messer" (which, by the way, means knife in German) and the date. Over the next three weeks, we received dozens of calls from workers explaining the emergency that had kept them from attending, and asking for a knife. We held the line, and attendance ceased to be a problem. Obviously, well-paid craft workers could afford to buy any knife we were giving them; the point truly is that the Messer pocketknives are a symbol of appreciation—and honest appreciation is always well taken.

In 1993, as we were working to grow the company in spite of the recession, the unions that represented our employees decided to draw the line on subcontracting. In spite of the facts that our direct craft force had been growing steadily and that large, sophisticated owners had made it very clear that if union employers like Messer could not guarantee their projects access to all of the performance resources in the region—both union and non-union—they would simply contract with non-union general contractors who could subcontract without constraint, the unions once again decided to target Messer as a way of enforcing the subcontracting clause.

This time the outcome was different. Within five days, 100 percent of Messer's foremen and 70 percent of Messer's craft workers resigned from their unions and returned to work. This was a very difficult decision for the workers and their families because many of them had been union members for many years. I believe that in the end the workers made the decision to work directly for Messer not for money (they were and are highly skilled and can command top dollar) and not for benefits (the health care and retirement that we provide directly is about on par with the union plans). In the end, I believe they came back for information—our sharing with them how they fit into our company's future and our giving them the information that allows them to make informed decisions about their own future.

Messer's path to steerageway required open sharing of all information with all employees.

The agenda of our craft meetings remains the same to this day. We share our jobs with them—what we as executives do day to day to earn our pay. We share the opportunities, challenges, and results of the company—so that they know what we are trying to accomplish and can help us get there. We share the same annual report that our ESOP participants receive, including our financial results. And, most important, we share our appreciation for their contributions to our success—recognizing their individual growth, their success in customer relationship building, and their support for our sales efforts. It works! A truly engaged work force drives progress, sets the tone for safety, and models the level of quality we expect in our projects.

GRASS ROOTS SALES
One strategy for engaging the craft workers in the sales process is to make sure that we respond when they give us leads. If a craft worker presents a project lead, someone from management has to

report back what we found when we followed up on the lead. This simple process creates real accountability and real value.

Some years ago a craft worker from southern Kentucky told his boss about a foundry that was being planned in London, Kentucky. His boss passed the lead on to Sales, a call was made, and the salesperson was told that it was just a rumor—that there was no real opportunity. Ordinarily, that would have been the end of the story; however, Messer had a requirement to report back to the craft worker. When the craft worker was told that there was no project, he responded, "How stupid can you be? My brother-in-law is on the planning commission that approved the plans last week!" Sales went back to work. Messer not only built that project but has built several more projects for what became a great, long-term customer.

In 2002, after more than ten years of ESOP experience, we asked the senior managers (by then there were more than fifty of us) what decisions mattered most in creating Messer's success.

The response was overwhelming. The decision that mattered most to Messer partners was our decision to put direct performance at the core of our value proposition—the decision to remain builders rather than becoming mere managers of the building process. Staying connected to the work—and to the workers—means that we have access to better information when we plan; it means that we have a better understanding of what we are asking others to do; and it means that we have an added level of Messer leadership on our projects.

Our Messer craft leaders model a commitment to quality, to safety, and to performance that sets the standard for every worker on our projects and elevates expectations across our markets. We are determined not to separate ourselves from our workforce. First, because if you separate yourself from those who drive value creation, you have no right to a forty-year planning horizon—understanding value creation in an intimate way is the key to sustainability. Second, and most important, we will not separate ourselves from the workers because they control the fun! They are out there every day turning dreams into reality at an epic scale. We are blessed to get to share in that fun, if we are connected to them, if we appreciate them, and if we support them.

Finally, there is this added benefit: every person has a circle of influence beyond the sight of his or her employer. Laborers teach Sunday school to bank presidents, and carpenters have daughters who lead corporations. Being accountable to workers and giving them a voice can be messy and uncomfortable; but it is among the highest value strategies that we have discovered.

Aisin Knoxville, Tennessee (top) and Aisin London, Kentucky. Projects brought to the table by a craft worker.

The Better Business Bureau International Torch Award

The Cincinnati BBB Torch Award

The Louisville BBB Torch Award

SERGEANTS MAJOR

Our intention in growing Messer has been to create an open marketplace for talent, where employees can enter at any level with the opportunity to grow to the highest level of leadership. This "drive for personal growth" caused us some challenges over the years because it sometimes made it look as if anyone who continued doing a job for a long period of time was simply not committed to growth. This is, of course, not the fact; the fact is that you must continue to grow to do any job well over a long period of time. Promotion and responsibility change are optional; growth is mandatory. In addition to this challenge of perceived growth, we also had the challenge of young, gifted hires coming into the company who had never had the benefit of deep exposure to craft work. Those young people needed real mentoring and guidance from seasoned craft professionals. The Messer solution was to create a new position with the title, "senior superintendent." These accomplished craft leaders are asked to act in the role of sergeants major in the army—to be the designated leader of the enlisted troops and the voice of the enlisted troops to the officers. In Messer, what this means is that the senior superintendents are the leaders in getting work done; but that they are also empowered to make the following statement to any member of the project team, "I will never have Strange's job, but you won't either if you don't start listening to me."

Messer's path to steerageway is based upon remaining closely connected to the work: to embrace the people, the processes, and the joy of building.

As we continued our growth, we had to face the reality of working in a very small market area. Even though our market share in Greater Cincinnati was rather small by P&G standards (in consumer goods, leaders strive to capture more than 30 percent of the market), the reality of the construction market is much different from selling consumer goods. If you are selling toothpaste, all customers who have the money to buy a tube are equally desirable. In construction, all customers are definitely not equal and all projects—and project types—definitely do not have the same risk. So even though we were doing less than 10 percent of the available work in the Cincinnati area, it quickly became evident that we could not achieve the growth goals in our business plan without creating new concentrations of opportunity in new markets.

Three influential Messer leaders:
Star Miller, Senior Superintendent; Kathy Daly, CFO; Bob Verst, Senior Superintendent

Ohio State University Biomedical Research Project Mat Pour: 150 Messer people from across the company placing 7,160 cubic yards of concrete—Messer's bggest pour.

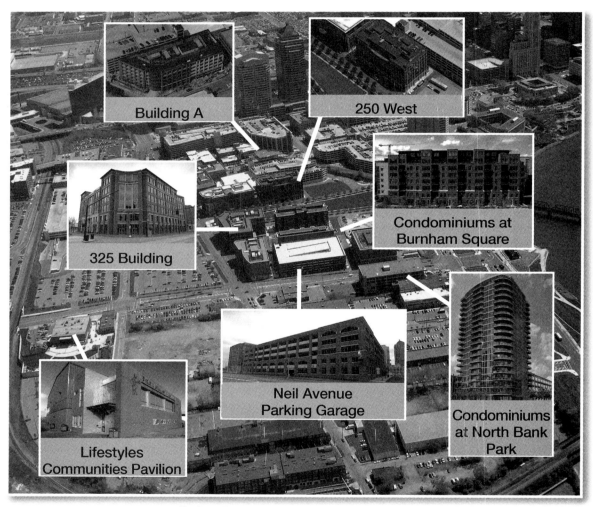

Building A

250 West

325 Building

Condominiums at
Burnham Square

Neil Avenue
Parking Garage

Lifestyles
Communities Pavilion

Condominiums
at North Bank
Park

Nationwide Arena District, Columbus, Ohio

Messer had, of course, been working in surrounding cities for its entire history—and had at one point followed customers across the eastern half of the U.S.—but our culture treated assignments out of town as, at worst, punishment and, at best, great suffering; so there was not a lot of enthusiasm for growing geographically. We needed to move from "sending" people out of town for a single project and move toward creating concentrations of Messer leaders in new cities. We needed leaders who would become deeply engaged in the local industry and in the local community, and who would build a circle of influence in support of Messer success.

There was good consensus that some folks needed to relocate on a long-term basis rather than just for a project—but there was also broad agreement that that relocation was for someone else, not for me. We needed a new model and, as happened so often in our history, a Messer leader stepped up to help create that model. That leader was Jim Hess. Jim was then and is now driven by challenges. A star athlete in high school, Jim saw the gap and drove forward full speed. He announced a date when he and his family would be permanently relocating to Columbus, Ohio; then he started recruiting star performers to join him. Being asked to join a leader that you respect in embarking upon a bold, new adventure is a fundamentally different experience from being assigned to "go there and get a job done." Other leaders quickly stepped forward to join Jim, and Messer was on its way to building a new, regional model for company growth.

THE MESSER MODEL

One of the positive outcomes of Jim's leadership was creation of the Messer Regional Model. When we moved from assigning leaders by the project to moving people for the long term, we were faced with having to identify what should be centralized in Cincinnati and what should be exported to individual regions.

The executive officers came together (very appropriately in Columbus, Ohio) to design the infrastructure for market leadership. At that time, we called it the "Fifty-Million Dollar Model" because about $50 million in revenue was required to carry the overhead we felt would be needed. Fifty million dollars of revenue in the smallest market we served at that time came to about 10 percent of the non-single-family, complex building construction work; so 10 percent market penetration as a goal flowed from the Fifty-Million Dollar Model. The model has evolved over time, but the fundamentals remain the same: a sales leader, a safety leader, a craft leader, an

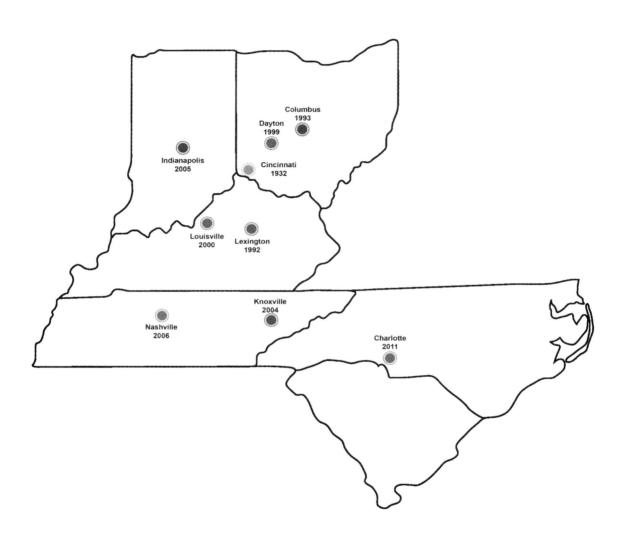

Regional map with the years the offices opened.

administrative leader, two operations leaders, and, of course, a market leader, who would need to be an officer so that he or she would be deeply engaged in company-wide planning and decision making. The Fifty-Million Dollar Model guided decisions about where we should go and about who should go. Today we are entering regions where we need to do many times $50 million in revenue if we are to be the market leader, and today we are clear that we should strive for much higher market penetration in our targeted market segments; however, the fundamentals for success—a company leader willing to relocate, targeting customers not projects, raising the bar on safety, self performance, and community leadership—remain at the heart of Messer growth.

Messer's path to steerageway is based upon individual leaders who are willing to take the risk of leading from in front.

Despite the ups and downs of the economy, Messer grew steadily. We had a well-developed career model for Operations employees that was very attractive to our core group of employees—young, aggressive, white males, just like us. Of course, there is a problem with that statement; not everyone in leadership was a white male. Mostly because Kathy Daly was one of the leadership team, we were forced to address several issues around diversity and inclusion. The most obvious was that we had almost no minorities on a path toward senior leadership and, even though Kathy was CFO, very few women. Less obvious, but just as impactful, was the separation between Operations (those who build buildings) and Administration (those who manage the business). Because we had grown up in a world where process was not supportive of performance, most of us who had come from an Operations background honored strong personalities and strong reactions. We had grown up leading those we respected and driving before us those who we did not. To say that Kathy Daly is a person who will not be bullied by a bunch of loud-mouthed construction guys is a real understatement—and led to some really interesting planning sessions—where both words and objects flew from time to time.

CHANGING THE WORLD

Back in 1970, job access was obtained primarily through the printed media—newspaper ads. If you would have looked at the help wanted ads at that time, you would have seen two different headings: "Help Wanted Male," which would list jobs like Management Trainee, Engineer, Accountant; and "Help Wanted Female," which would have

Messer Partners in 1999. The tenth anniversary of employee ownership.

listed jobs like Gal Friday, Barmaid, and Hairdresser. In June of 1970, Procter & Gamble hired Messer to build an expansion, and P&G struck a small blow for diversity by requiring Messer to advertise for its project accountant under a third heading—"Help Wanted Male or Female." The week that ad ran there happened to be a young accountant named Kathy Daly, two years out of the University of Dayton, who wanted to get back to Cincinnati to help her widowed mother. When she looked in the paper, she had an opportunity set of just one job. She got the job and in time rose to be Messer's CFO. That small act of leadership by P&G changed Messer's world. We are a different company, and I am a different person because of Kathy Daly's quality as a person, her quality as a leader, and, yes, her perspective as a woman.

Messer's path to steerageway requires that we win the true competition. The true competition is not for the next construction project; the true competition is for the next generation of talent—which will include folks from both sexes, and every race, ethnicity, and cultural background.

Over time, Kathy and some very smart people who work with her taught us that value can be measured in many more ways than just in cubic yards of material moved. It turned out that we had smart people all over the company who could contribute more, if they were allowed into the conversations and

Kentucky Exposition Center, 2007

Kentucky Department of Transportation, Frankfort, Kentucky, 2004

University of Kentucky Center for Pharmaceutical Science and Technology, 2006

Fleming County Replacement Hospital, 2008

BUILD KENTUCKY AWARD WINNERS

if their contributions were valued and respected. These "support" people actually have the vocabulary and business skills to help us Operations people understand our customers' business plans—a key element in moving from serving projects toward partnering with customers. What we were so proud of knowing about our relationships with craft workers we had never applied to our relationships with accounting, estimating, IT, etc. Creating a strong core of solid business practices became as important to our growth as our ability to drive field production.

A COMPELLING METRIC

The data tells us that many construction companies actually go out of business making a profit. They are able to estimate and execute the work, but they have such poor business practices that they never collect for the work—and they die from poor cash flow while they are arguing with their customers. Some years ago, Tom Keckeis and I made a trip to Florida to meet the leader of our new surety provider. Messer had been a customer of USF&G for many years; but through a series of mergers and acquisitions, we found ourselves as the customer of Travelers, a much larger and more disciplined organization. We were (and are) a good account, so there was no question that we would continue to earn surety credit to support our work. Still, we had been used to dealing with people we had known for years and who understood our story.

So we needed a way to share, in a short introductory meeting, the "Messer Difference." We took one piece of paper with us and handed it to the Travelers' surety leader. It was an aged receivables report with just two lines in big print. The first said, "Total Revenue—$500,000,000"; the second line said, "Receivables over 60 Days—$160,000." We made our point.

Messer's path to steerageway includes every employee being in the game. We are all in the same box: one set of expectations, one level of accountability, and a shared commitment to growth.

Committing to growth based upon project leadership capacity rather than market opportunities led to an obvious conclusion: we must invest in processes that accelerate the growth of project leaders. We tried a number of outside programs in support of that growth (and still value and use outside learning experiences), but we came to the conclusion that internal programs create all sorts of synergy that just cannot be achieved with outside classes.

The benefits of mixing people from different regions and different departments, so that they get to know and respect each other; the energy that comes from driving a concept across the company, so that we all share the new vocabulary around processes and strategies; and the power of having a complete feedback loop, that allows us to take the classroom lessons directly into the field and then pass back learnings for future classes, all drove us to find an internal solution for professional development.

As usual, we had the right resource already in the company. Bill Krausen came to the company from Western Kentucky University and worked his way up through Estimating to lead that department. But Bill had another love; from very early in his career Bill embraced teaching, serving as an adjunct professor at local universities. We had a need and we had a leader. When we proposed that we bring the two together, Bill embraced the opportunity. Over the following twenty years, Bill designed and delivered programs on a wide range of subjects, but his greatest contribution is on the topic of leadership. The Messer leadership program is a model for bringing diverse people together around a solid core of actionable learnings, but just as important, bringing people together in an environment that supports risk taking and builds teamwork. In their spare time Bill and Charlie Cook, a contractor from Philadelphia, created what has become a very successful leadership program for the Associated General Contractors of America.

EVERY EMPLOYEE A TEACHER

When I was young and still knew damn near everything, I had a project at East Bend Power Station called the Lime Unloader Facility. The project included a large round structure in the Ohio River that was connected to the shore by a concrete tunnel. The concrete tunnel spanned about a hundred feet and for some design reason had to be placed as a single concrete pour—meaning that the base, walls, and roof would all be formed and poured at the same time. This was a lot of concrete and a lot of load. I could design a temporary structure to support the load, but I was challenged by what to do to remove the wooden formwork after the load was applied. The two obvious choices—either beating the wooden formwork into splinters or cutting heavily loaded members—seemed likely to get someone hurt . . . and then I had an idea. I designed an interconnected system of low-clearance hydraulic jacks that would support key load points and could be lowered to relieve the load.

Argosy Casino, Lawrenceburg, Indiana

Hollywood Casino, Lawrenceburg, Indiana

Horseshoe Casino, Cincinnati, Ohio

PLAYING SUCCESSFULLY IN A NEW MARKET SEGMENT.

I was so proud of my inventiveness and design that I went looking for someone to show it off to. I went to Harry Ahlstrom, the project manager, unrolled my drawings and described to him in great detail what I had planned. Harry was very complimentary; told me that he looked forward to building such a neat approach; and then said that he would probably have just used sand jacks. I had never heard of sand jacks so I asked what they might be. Harry said that they were just a dumb carpenter's approach and not nearly as neat as what I had come up with. When I asked again, he sketched out a simple assembly with a short piece of pipe filled with sand to carry the load; and to release the load, you just cut a hole in the side of the pipe and pulled out the sand.

My "smart" approach would have cost tens of thousands of dollars; Harry's "dumb" approach cost a few hundred.

The best resources we have for learning are the creative, committed people who have been out there on the line.

In 1997, Messer's revenue jumped from $240 million to over $300 million as the result of the very aggressive schedule on the Argosy Casino Project in Lawrenceburg, Indiana. Everyone was stretched, but we proved we had the capacity to manage that challenging project, along with our normal book of work. The measure of our discipline around the capacity model is that we planned toward lower revenue in 1998 to make sure that our capacity and our support processes had time to catch up with our growth. It is not very comfortable to project declines in revenue, but it is very comforting as a leader to know that you are growing based upon a solid foundation of leadership capacity.

Messer's path to steerageway is founded upon supporting personal and professional growth at the individual employee level.

As Kathy and many others sought to create value in places other than Operations and as we worked to keep our focus across a growing number of regions and project sites, we came to understand the difference between participation and ownership. For every human being, the level of engagement and commitment can vary widely. At the lowest end of the spectrum there is grudging compliance, that which is done from fear of retribution or punishment. In many employment situations that is the norm. As employers work to communicate with their employees the level of engagement can move all the

133

The Christ Hospital Courtyard Project, 1988.

The Christ Hospital Medical Office Building, 1994.

way to informed support, that which is done with understanding and energy. This is what most employers dream of and aspire to.

We have learned that there is a level of engagement far beyond informed support: it is ownership, embracing one's individual responsibility to create a successful outcome. The difference between informed support and ownership lies in engaging the individual's emotional intelligence. An employee who understands the plan and pledges to work the plan as hard as he or she can is a good employee; a person who understands the plan, and then uses all of her or his intellect, creativity, and energy to improve the plan during implementation is an owner.

Everybody talks about empowerment, but what they usually mean is that an employee is free to support management's plan in flexible ways. The path to ownership is a form of empowerment that frees the employee to say to management, "This won't work for me." Or, even more daring, to simply change the plan to make it work better, with the expectation that he or she will receive support and recognition for the innovation. This is scary stuff for decision makers and requires a super solid core of values and value definition so that we are not straying all over the place, but remain aligned around customer service and engagement. We have found that both our ESOP participant owners (Messer salaried employees) and our craft workers (who have a separate cash retirement plan) are equally willing to embrace ownership if they know that they have our support and our trust.

OWNING VALUE CREATION

Messer has had an interesting and informative relationship with The Christ Hospital, one of the biggest and among the best-run hospitals in Cincinnati. In the mid-sixties—in the days when projects, not customers, were the focus of our efforts—Messer performed a project for the hospital that resulted in a dispute. I have no idea who was right because I wasn't around at the time, but we all learned who was in charge: the customer, who for the following twenty years simply elected not to consider Messer for its work.

Finally, after twenty years we were able to re-engage, with the award of a large clinical project to Messer. We were determined to do a great job on the project, but the act that solidified our relationship had nothing to do with our scope of work. One day I got a call from the facilities VP at Christ, who started the conversation with, "Well, you've gone around me, and there is nothing I can do about it." I was understandably concerned and asked what had happened.

Here is his story:

"For several months I have been trying to get the city to fix a big pothole in the public street right in front of the hospital. Every day my boss asks me when it will be fixed, and I have to make excuses about city rules and the bureaucracy. Today the hospital president came to work to find a pickup truck parked in the street and a worker, with a Messer jacket on, tamping cold patch into the pothole. When the president asked how the city had come to approve the fix, the worker replied, 'Oh, I didn't ask anybody's permission. I am a carpenter with Messer and I figured there shouldn't be a pothole in front of your hospital, so I bought some cold patch and I'm fixing it.'"

"So," the VP finished up, "now my boss thinks Messer is a company that has workers that care so much that they can solve his problems without asking permission, and I can't get rid of you even if I want to."

This is the most powerful sales story I know, but I know a lot of others that are awfully close—where Messer employees took action to create value for customers and the customers came to the correct conclusion that a company where workers are empowered to add value without asking permission is a good partner to have.

Messer's path to steerageway requires that every Messer employee be an owner and a leader, with both the willingness and the ability to influence events so that goals are met.

I close this chapter with what I consider to be the lever that moved people to action in support of growth and in support of each other. It is a concept that we call "Mutuality," and I share some thinking that I captured back in 1998, when we were wrestling with the question of what we could expect from each other.

September 28, 1998
MEMO TO: MESSER PARTNERS
Re: THE CONCEPT OF MUTUALITY
    (One of the needs of the Messer Partnership)

Mutuality—condition or quality of being mutual; reciprocity; mutual dependence.
Mutual—possessed, experienced, performed, etc., by the members of a group; reciprocal.

An enduring relationship. The Heart Center of Greater Cincinnati at the Christ Hospital, 2004...

...and, The Orthopaedic and Spine Center at The Christ Hospital, 2012.

Several years ago the officers of Messer came together to address the roles and goals of their positions. The reason for the meeting was a growing feeling that the individual officers were not able to grow in their new jobs and contribute to the success of the company because they did not have a clear understanding of their responsibilities and because they did not have a clear feeling of mutual support. Too much time was being spent in meetings, preparing for meetings, trying to find out what was happening in meetings, and creating explanations for others, especially the CEO, to make sure no one felt left out or neglected. In short, six extremely gifted individuals were under-producing because they lacked mutuality—a set of shared goals and a commitment to group success.

Over the course of two days, the officers analyzed what the job of officer means distinct from other responsibilities in the company; what the requirements for success as an officer are; how the job is perceived by the rest of the company; and what each of them would have to do to assure success of the others. This last concept turned out to be the critical issue. Since the rest of the company is extremely sensitive to the officers' actions and statements, it was clear that any sort of public fussing sapped strength from the important business at hand. In addition, the fear of second-guessing or public disagreement was keeping individual officers from taking risks and implementing needed changes. The result of that meeting was a compact among the officers:

- The CEO's job is not to make sure everyone else is doing his or her job. He has distinct responsibilities separate from being the foreman of the Officers.
- Debate in private; support in public. The Officers will present a unified face to the rest of the world. If there is sufficient disagreement over a subject that even one officer cannot support the plan in public, then the matter is not ready for action.
- Those who are present will make decisions about the matter at hand. Once the decision is made it is a Messer decision and will be supported. If it needs to be changed, the change will be a group decision.

This compact has allowed the individual officers the room to grow as individuals and the ability to take the needed risks to lead in the growth of Messer. The feelings of mutual trust, mutual support, and mutual respect that it has engendered have been very powerful. In addition, contrary to intuition, this approach has resulted in a feeling of being better informed on what the members of the group are doing.

It is my opinion that the current state of the relationship among the Partners of Messer is analogous to the challenges faced by the Officers. We have many people who are concerned about how to contribute; we have many people who are concerned about how they are perceived, especially when they take risks; we are evolving job descriptions as we grow so there are a lot of risks to take; and there is a growing feeling of not being well-informed despite the number of meetings we attend. (And, too much time is being spent preparing for meetings, attending meetings, and following up on meetings!) What is missing is a compact: a compact pledging mutuality.

What would this compact say? Some examples might be:
- I trust my partners; I will go out of my way to make sure that they and others know this.
- I am committed to my partners' success. I will communicate this commitment as often as possible and I will seek ways to assure that my partners succeed.
- I realize that my partners care what I think. I will take an active interest in what they do and will seek ways to support their projects.
- Partnership is active, not passive.
- The growth of Messer depends upon the growth of my Partners; I will find ways to support that growth.

What would this mean in real life?
1. Using our relationships within the company in support of our partners' success:
   - Taking a more active interest in professional development.
   - Actively supporting our partners in conversations within the company.
   - Going out of our way to emphasize the contributions partners are making and asking others for help in support of their success.
   - Supporting partners' actions and initiatives, especially when there is risk and especially when there is failure.

2. Using our relationships outside the company in support of our partners' success:
   - Influencing subcontractors to perform for our partners.
   - Introducing partners to business partners who can help them meet goals.
   - Looking for opportunities to publicly support the partnership.

Disagreeing in private but being mutually supportive in public. (It really is us against the world!)

This is not a call for mindless support or some kind of group hug. This is recognition of the fact that success breeds success and that the rest of the world is trying to figure out where we are coming from and where we are going to as a company. Our mutual support can have a significant impact upon the world's view of Messer. A simple example from my recent experience makes the case: A large subcontractor whom I met at a social function cautioned me about our fast growth and the fact that it was causing us to put young, inexperienced people in charge of large complex projects. My natural response was to get into his problems and to assume that we were not performing up to my expectations. He probably left feeling his concern was justified as a result of the questions I asked about his bad jobs with Messer.

We are growing, and we do have a lot of young people, but every one of those young people is supported by one of my partners. I could have asked him what jobs he has been involved with; told him how much confidence I have in the partners who are responsible for performance on those projects; told him how sure I am that those partners would respond to any concerns he may have; and offered to facilitate the follow-up by bringing his concerns to their attention. If the concerns are valid, of course they should be addressed; however, the opportunity I passed was to establish an expectation for success. I could have helped my partners grow through my support—the sub respects me; he will elevate his respect for those I overtly trust and support.

This may seem to be a problem specific to me because of my title and personality, but I think it gets at something I have been feeling in our Partner's Meetings. There seems to be a general agreement that we like and respect each other but that individual success is an individual responsibility. The concept here is for each of us to take responsibility in a much more proactive way for the success of our partners. From the outside we should look seamless—not all of equal ability and experience, and certainly not all the same, but all committed to mutual success. From the inside we should minimize the energy wasted in checking up on each other, questioning authority or commitment, and hand wringing. Each of us should personally take a much more active interest in the success of others.

This may not, at first, seem like a big change, but for those who may be struggling to meet our expectations, the impact of overt support will be tremendous. What would happen, for instance, if those who have significant influence with leadership subcontractors made it clear that they are paying personal attention to how those subs perform for one of the new project executives? The sub's performance will improve! Whenever we move off the sidelines and indicate a personal interest, the world notices. If each SPE as part of his or her buying of a job asks who else in the partnership the vendor is working for and indicates a real interest in assuring success of those projects, it will be noticed.

Finally, there is the matter of getting past our differences. Because we have different abilities, different styles, and approach our jobs in different ways we have concerns about the performance of others. We cannot allow these concerns to turn us into neutrals or non-supporters of those partners who are not like us. This problem of variation in style is dealt with by the officers because we are so close personally and because we must deal with each other almost every day. For partners in different markets or in different departments, it is far easier to simply become passive toward each other. The result can be low trust and little interaction. We can't learn from each other and grow if we don't have an active relationship. The active strategy for getting past these differences is a new definition of success which makes those who we perceive to be at-risk our personal responsibility—not to make them over in our image but to support their style however we can to assure that it works. We must commit to mutuality with our partners!

Thanks for your support,
PSS

> **Mutuality, that commitment to active engagement and trust, is the heart of Messer strategy for finding steerageway.**

Messer

**An example of developing mutuality with a customer:
Cincinnati Children's Hospital Medical Center (1988–2012)**

142

Substructure and Superstructure—Unit 2, East Bend Station, Rabbit Hash, Kentucky
Cincinnati Gas & Electric Company

Substructure—Unit 1, East Bend Station
Cincinnati Gas & Electric Company

Substructure—Units 7 and 8, Superstructure Unit 6,
Miami Fort Station, North Bend, Ohio, Cincinnati Gas & Electric Company

Coal Handling Foundations, Waste Water Treatment & Warehouse,
J.M. Stuart Station, Aberdeen, Ohio   Dayton Power & Light Company

Sitework and Substructure—Unit 2, Spurlock Station, Charleston Bottoms, Kentucky
East Kentucky Power Cooperative

Substructure—Unit 4, Superstructure—Unit 5, Substructure and Superstructure—Unit 6,
Walter C. Beckjord Station, New Richmond, Ohio   Cincinnati Gas & Electric Company

Power plant projects

8

CLOSING THE GAPS

The Players
TOM KECKEIS
THE EXECUTIVE LEADERSHIP TEAM
YOU

Culture eats strategy for lunch!

CLOSING THE GAP BETWEEN STRATEGY AND REALITY

I read a lot of business plans. Most of them scare me to death. They are so strategic, so focused, so goal driven, and so metric supported that I fear we will never be able to compete with such smart people.

But I have noticed over the years that not all of these smart enterprises thrive, or even survive. What I have concluded is that intellectual intent and clear articulation are good assets, but success depends upon closing the gap that always exists between all that good thinking and articulation and the reality of how people live their lives at the point of value creation. In fact, the gaps are all around us. There is a gap between customer perceptions of what we do and how we think and feel about what we do. There is a gap between the processes we have in place to serve subcontractors and vendors and how it feels to work for Messer on a project. There is a gap between being a good company and the community's perceptions of what it takes to be a good corporate citizen. Most important, there is a gap between our individual responsibilities and expectations and the feelings and expectations of those we touch through our leadership.

At Messer, this last gap is most clearly seen in the separation between production—the Operations people in the field—and support—the many people who do the blocking and tackling required to keep the enterprise healthy and vibrant. Closing the gaps—not just good strategy and good articulation—is the most important job of company leadership. Going forward, strategy will matter, structure will matter, discipline will matter; but the strength of Messer's culture around closing the gaps will determine what Messer will become.

EXPLORING THE GAPS

An experience that stands out in my mind is the day I first wore a suit and tie to a construction project. I was leading a project at Miami Fort Power Station and was having trouble getting the design engineers to sign off on my pour plan for placing five thousand cubic yards of concrete in a large stack foundation. After three meetings with no resolution, I went to my boss and told him that I was at a loss. I had a good plan, I had addressed their concerns, but they kept coming up with new questions rather than releasing the pour. He asked, "What do they look like?" I answered, "Just a bunch of stuffed white shirts with ties on." He then said, "Before I get involved, do me a favor. Go to one more meeting to present your plan; and you wear a white shirt and a tie." It was great advice and it worked perfectly (it closed the gap), which would be enough reason to make the day memorable. But here is the rest of the story.

As I came through the project gate wearing my brand new suit and tie, I had the bad luck to have the first worker who saw me be my cousin Stanley. Stanley was a millwright, a brilliant, talented craftsman who moved at the speed of cold maple syrup. Here he came with that steady ground-respecting walk of his. He got right up into my face, about two inches away, peering at me as if I was some new specimen that he had never seen before and said, "I thought it was you, Petey. At this rate, in one more year of college you just won't be worth a damn, will ya?" This comment still rings for me when I meet college-educated people who have the attitude that their degree demands respect. (A gap was developing between Stanley and me.)

But the most memorable moment of the day came when I completed my meeting and was making my way off the site. I thought I would just walk past a project we were doing in the kinghole, the large and very deep steel piling coffer that extended below the river to house the cooling water intake. I walked out over the kinghole and looked down through the steam and mist to see our people working below. John Clark, the carpenter foreman, looked up, saw me, and went into what appeared to me to be hysterics. He was jumping up and down, waving his hands and clearly yelling something—although there was far too much plant noise for me to make out words. Fearing that some disaster was happening while I watched, I took off my new suit coat, climbed down fifty feet of ladders, and made my way through ankle deep muck to get to John. Out of breath and covered with mud, I yelled, "John! What's wrong? What's wrong?" He looked at me with a grin and said, "Not a damn thing now that you are down here where you expect us to work." (John had just closed the gap.)

CLOSING THE GAP BETWEEN STRUCTURE AND PURPOSE

I am a big fan of the unconstrained data point. If you have just one data point, you can move it to wherever you want to start, and you can draw a line through it to wherever you want to go. Multiple data points make life messy; they lock you into a starting point and channel you toward a future that is not as flexible as you could wish.

The ESOP structure is a very powerful financial framework. When taken as a single data point, it can be used to facilitate ownership transfer, raise capital, and reward some participants. It is the introduction of the rest of the data that creates the complications. An ESOP can also be about giving employees a voice, creating a deeper level of engagement, and empowering growth through a personal connection to capitalism. An ESOP can provide the ultimate balance to the inequity of the employment contract. But to accomplish its potential the ESOP cannot stand alone; it can't be that unconstrained data point that can be moved around at will. A truly effective ESOP must be anchored to some fundamental purpose.

For Messer, the fundamental purpose that anchors our ESOP is: "To allow competent, caring, committed human beings to make the trade that matters most in life; to trade their talents, efforts, and commitment for value—value for

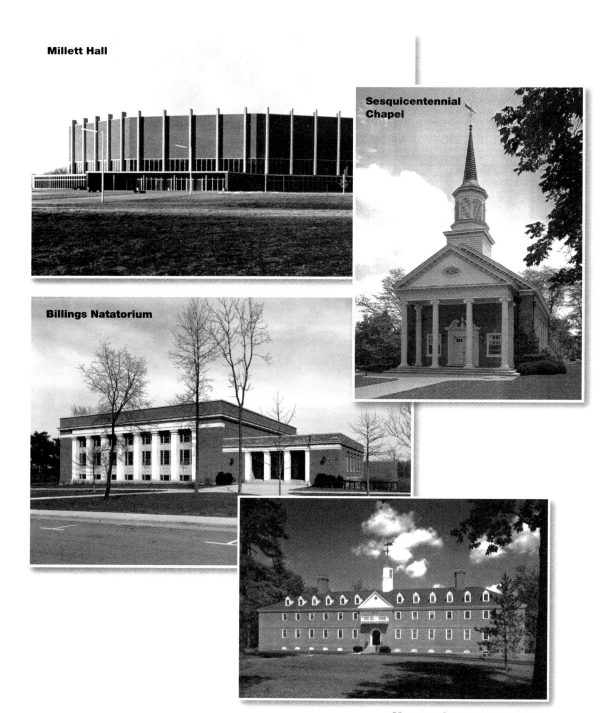

Millett Hall

Sesquicentennial Chapel

Billings Natatorium

Marcum Conference Center

MIAMI UNIVERSITY: ANOTHER ENDURING RELATIONSHIP.

themselves, value for their families, value for their customers, value for their communities, and value for their profession."

The interesting thing about our purpose is that it does not include the path that we will follow. I love building, it's a wonderful profession, and it is all I have ever done, but building is not the fundamental purpose of Messer, it is a path to our purpose. Over our forty-year planning horizon, I cannot predict all of the paths we will find to our purpose, but I am absolutely certain that if future leaders remain focused on our fundamental purpose it will be an exciting and rewarding journey.

MY DESCRIPTORS

When I am asked to describe Messer, I generally respond with three statements:

- Messer is a get-rich-slow scheme.
- Messer will remain vibrant as long as aspiring leaders believe that Messer is slightly broke and that it is their job to fix it.
- We are just in charge until the grown-ups come back, so let's try to have some fun.

CLOSING THE GAP BETWEEN COMMUNICATION AND CONNECTION

We live in the communication age. I can reach hundreds of people with one click of my mouse, and I can create messages that can be retained and played back thousands of times. We are so blessed with tools that we sometimes mistake using the tools for creating a result. Communication is not what we send; communication is the portion of what we send that others receive and act upon.

I have a simple test for communication. Look for action: intellectual action, emotional action, or physical action. If there is no action, you have not communicated, no matter how good your intentions and your tools.

TO COMMUNICATE TO ME IS NOT TO KNOW ME

One of my attitudes is that rodbusting—putting the steel bars in place that will be surrounded by concrete—is among the worst jobs in construction. Those steel bars weigh 550 pounds per cubic foot; it's hot in the summer; cold in the winter; and much of the activity takes place at the level of your feet, so you have to bend over a lot.

P&G Parking Garage, 1970

P&G Office

P&G General Office, 1955

**P&G General Office
Addition, 1970**

None of that, however, is what makes me think it is the worst job. I think rodbusting is the worst job in construction because everything you do gets covered up; you can never point to your personal work with pride. With this attitude, it is not surprising that I came to some wrong conclusions about rodbusters early in my career. I concluded that they were people with strong backs and a capacity for boredom, and that they needed someone like me to communicate to them regularly about the plan and purpose of their work.

My attitude changed when John Gressler retired. As John's last day approached, I became concerned that I wouldn't be there to communicate plans and purposes after his retirement, so I thought I had better talk with him. You need to picture a twenty-five-year-old know-it-all talking with a sixty-five-year-old construction worker. I earnestly and condescendingly asked John about his plans, and after he appropriately told me it was none of my business, he decided to tell me anyway. Here is what he said:

"You don't deserve an answer, but I am going to give you one. There are books I am going to read that I haven't read—because when you bust rods all day you are too tired to read at night. There are places I am going to go that I haven't been—because I sell my services by the hour, and if I had missed a day of work I would have missed a day of pay. And, I am going to spend more time with my stamps."

It turned out that this man, who I thought I was leading with my great communication, was an expert on stamps—not postage stamps, but Revolutionary War tax stamps. He was regularly consulted by other experts on matters of authenticity and value, and he had a great collection, which he shared with me—after I stopped communicating and showed an interest in connecting.

In John, I had access to a brilliant mind; and until the day he retired, I had been so busy communicating with him that I never asked him a question.

While we are busy enjoying our tools—creating agendas, sending e-mails, presenting PowerPoint presentations, and giving speeches—we would do well to think about how many Johns there are in our lives. Communicating about a topic does not create a connection; and communicating about a topic certainly doesn't establish ownership of the topic. Only through showing an honest, personal interest in others can we create connections; and only through teaching the topic to others can we come to ownership. This is one of many instances where we are increased as human beings through giving freely of ourselves.

CAMP JOY

I have mentioned that the twenty employee investors went on a weekend retreat to Camp Joy shortly after we completed the transaction to buy Messer. Camp Joy is an experiential learning center just north of Cincinnati. On our retreat we shared bonding experiences and did some really focused planning. One of my great learnings from that experience was about connections. I learned the power of capturing the moment. When I set up the meeting, the facilitator asked if I wanted to pay an extra $250 to have a videographer film the experience. My first answer was, "No one will ever watch the video." But after some conversation I agreed to spend the money. For many years, a portion of that video was played at every Messer orientation session. The activity being filmed was a path-finding initiative, using compasses and instructions to arrive at an end point. This is something that we engineers should have been good at, but we were so busy "taking charge" that we had trouble following the instructions. At one point, the camera is squarely on me and the videographer says, "This is your camera man. I am never supposed to speak while filming, but I need to observe that this is the most lost individual that I have ever seen." Whatever gap that might exist between my title and accomplishments and the reality of a new hire pretty much evaporates as we laugh together while watching that video.

CLOSING THE GAP BETWEEN EXPERIENCE AND LEARNING

Experience—doing the same thing day after day—wears us down, while learning to do old things better and new things well builds us up. It is a sad fact that if my grandpa were to come back to life today and walk on just about any construction project with his carpenter tools, he would feel right at home. The sorts of tasks, tools, and materials that we are about today are not much different from what he would have seen over a career that spanned

Indiana University CIB Data Center

Indianapolis Children's Museum

Scioto Mile, Columbus, Ohio

**Indiana University
Innovation Center**

MORE AWARD WINNERS

from 1913 to 1975. Oh sure, we use more power tools, have better hoisting equipment, and we are a lot safer; but when you compare our construction projects with any other work environment—think of the changes in computing, in business, in communications, and in medicine, where a practitioner from 1950 couldn't even recognize the tools or the processes—we are a triumph of experience and energy over learning.

Over the past century, the construction industry has embraced the perfect defense against innovation. When times were good, we were too busy to innovate; when times were bad, we were too poor to innovate. In the twenty-first century, part of the new normal for construction is that we can no longer be the industry that does more with more. We are headed toward a stratification of the marketplace. Those construction companies that embrace innovation and are willing to invest their own resources in new ideas will find strategic opportunities with sophisticated customers; those construction companies that remain attached to what made them successful in the past will live in an ever-more commoditized market.

TAKE TIME TO THINK

Back in 1926, Harvey Firestone wrote a book titled *Men and Rubber, the Story of Business.* Most of what Mr. Firestone shared is a complete anachronism today; but he included one timeless piece of advice: "Take time to think."

The gap between experience and learning is bridged by thinking—thinking before we act about how we can be more effective, and thinking after we act about what we have learned. Here are the most important learnings I have acquired over twenty years of being a CEO.

- Good ideas don't come from job titles. Good ideas come from human beings: human beings who understand the goals; human beings who understand their role in achieving the goals; and human beings who believe that it is okay for them to have a good idea—even if it is different from what the boss just said.
- Caring is not a product of authority. We have people across Messer—from clerks to typists, from project managers to the president—who are living and dying every day with the successes and failures of the enterprise around them. Every one of those people deserves to be respected for the emotional commitment that he or she is making to the work.

- The most powerful strategic advantage we can have in the marketplace can be boiled down to one word. That word is trust. We live in a complicated world. If we have trust and things go wrong, as they surely will, we don't have to waste time and energy trying to find out who was wrong-hearted, wrong-headed, didn't care, or didn't try. We can get right to solving the problem and back to the important work of creating value for our customers and for ourselves. If we have trust and things go right, as we hope they will, we don't have to waste our time and energy worrying about: "Where do I stand?" "How do I look?" "Who's getting ahead here?" We can get right to the celebration, then back to the important work of creating value for our customers and ourselves. If we can maintain trust in our enterprise, we will always have a higher energy budget than our competitors. If we can export trust into our relationships with our customers—our building owner customers, our designer customers, and our subcontractor customers—we can change the world. I believe that trust is to human performance what water is to human health. If you get a lot of trust everyday, you might thrive. If you are deprived of trust, I really don't care how smart your business plan looks on paper, you do not have a sustainable enterprise.

CLOSING THE GAP BETWEEN TALENT AND OPPORTUNITY

One of the magnificent opportunities of construction is that there is no pre-scribed career path: no protective credentialing, no threshold of capital or equipment, and except for the most rudimentary sorts of tests, no bureau-cratic registration. I have known successful contractors from every conceivable background, and I have met CEOs in our industry who have followed incredible career paths. All of this begs the question, "What does it take to succeed in this business?"

Some years ago, I chaired a task force for the Associated General Con-tractors of America with members from across the country. We had been charged to speak more clearly on the question, "What does industry need from academe?" With the help of some gifted researchers, we gathered in-formation through interviews, town hall meetings, and surveys. We worked to make sure that we heard the voices of people at every level on the career ladder, and we spent many hours reviewing, sorting, and analyzing the data.

The results were both clear and compelling. The construction industry needs future leaders who have a foundation of sound technical understanding, but in addition, leadership success requires four elements:

- Oral communication skills. Oral communication is fundamental to translating concepts into actionable work packages and motivating others to provide the energy to get the work done.
- Written communication skills. The written word is still humankind's greatest invention for transmitting ideas and information across time and distance.
- Planning skills. Organizing, prioritizing, and recognizing dependencies are requirements for undertaking large, complex projects.
- Leadership skills. The willingness and ability to influence events so that goals are met is the heart of project management.

The sad part of the task force story is that in doing my research, I came across some work that had been done on the same topic by a PhD candidate some fifteen years before. His findings, gleaned from broad study of success and failure in the industry, were exactly the same as ours—that industry leaders, in addition to sound technical preparation, need a command of oral communications, written communications, planning, and leadership. What makes this sad is that fifteen years after he published his findings, academe had made almost no adjustments to accommodate and develop soft skills.

Sitting in the task force meetings after we had the data, when we were trying to decide what to do next, it was easy to understand why so little progress had been made. Most of the contractor leaders around the table were very self-conscious about their own skills in some of these areas. They compensated by openly avowing, "They had made it without all of that business school stuff." My personal conclusion is that we have a collective inferiority complex about these skills: not a good starting point for driving change. And, we are afraid of people who are not like us—a horrible affliction for those who desperately need new ideas and new perspectives.

SCIENCE VERSUS ART

In one of these conversations, a construction CEO accused me of being born with the gift of gab, and he said that oral communication skills were something you either had or didn't—they couldn't be taught. At that point, Charlie Cook, a CEO from Philadelphia, spoke up. Charlie is one of those construction leaders who took a different path. Charlie never wanted to join his father in the construction business, and since he had an older brother, he felt free to pursue his dreams. Charlie ended up running the business as a result of his dad's death and his brother's illness, and he brought to the job some truly unique preparation—he has a PhD in speech and theater arts. So at our meeting, Charlie shared that a few years ago a fellow named Aristotle had set forth these foundational elements of oral communication:

Ethos—establishing the right to speak to the matter.
Logos—the efficacy and cogency of one's arguments.
Pathos—investing the proper amount of emotion into the delivery.

We are hiding in the past when we think skills must be discovered rather than developed. We close the gaps when we search for personal capacity, then we support accomplishment. Every one of the needed skill sets—even leadership—is more science than art and can be learned and improved by motivated individuals in an environment of support.

The job of leadership is to foster diversity of thought and to help motivated people grow.

CLOSING THE GAP WITH THE FUTURE

I am very proud of Messer's career-length planning horizon. Allowing employees to have a voice in changing the place where they work, rather than requiring them to follow the prevailing wisdom of changing employers to find personal growth, is a true differentiator. As with most pronouncements by CEOs, however, there is a little bit of absurdity in a sixty-three-year-old guy embracing a plan for the next forty years. I am fond of saying to other construction company CEOs, who believe in their hearts that their company cannot live without them, "Don't worry about making plans; God already has a succession plan for you." It is part of my obsessive nature to not allow events to evolve, but to drive the action through planning and energy. So now I come to the part of this story that is about modeling transitions into the future.

I did not come to Messer with a grand plan to lead, to change the ownership structure, or to create great wealth. I came with a love for building and an honest admiration for those who build well. As a result of that love—and a number of less admirable personality traits—my job and my goals have evolved over time.

When we bought the company, we were distinguished by our obligations rather than our assets. As we grew and created assets, I tried to grow with the job, evolving my own skill sets, becoming better at engaging others, and working to develop some modest skill at listening. Somewhere along the way, I began to obsess less about what I could accomplish and to think more about what might be accomplished after I am gone. Frankly, that is a hard subject for one who is in love with his work to tackle. I have known a lot of people who see work as an obligation. Those people have no problem separating from work and seem to honestly look forward to retirement as a fulfillment of their efforts and dreams.

I have known a lot of people who have a somewhat different view of work. I believe that living in an environment of expectations, accountability, shared purpose, and accomplishment is one of the most joyous opportunities I can imagine. I have no expectation that I will be a happier, better adjusted, more fulfilled human being at some future event called retirement. I have not worked in preparation for a good life: work—honest, productive, effective work—is a great life. To quote my friend Sky Smith: "I have lived my life saying, 'Thank God it's Monday.'"

Which brings me to the subject of succession and the next chapter in Messer's history. I need to start by sharing Tom Keckeis's story. Tom grew up in construction, working for his father at the Edward T. Honnert Company—the oldest and one of the most respected general contractors in Cincinnati. Tom attended the University of Cincinnati in civil engineering, co-oped with J.A. Jones (a great company that grew its way to failure) on a civil works project and joined Messer at graduation. Tom is a gifted builder, a strong person, and an able leader.

None of this, however, explains how two people who are so different could become such good friends. Tom acts while I am talking. Tom sorts ideas down to the important few, while I never met an idea I didn't fall in love with. Tom is analytical while I am emotional. Despite all of these differences, Tom and I come together around the joy of building. Tom, because he is a different person than me and because he is strong enough to challenge me, has been filling the gaps for me for three decades.

THE SUCCESS OF A FAILED PLAN

Larry Keckeis spent his entire life in construction, starting as a carpenter apprentice and working his way to ownership. He loved building, but he didn't love the building business—the stress, the aggravation, and the long hours away from his family. One of Larry's goals was to find a more wholesome career path for his sons. He

Tom Keckeis

The real Pete Strange at fifteen in his dirty work clothes.

Afterword

LUCKY CURRENTS

Although the title of the book is *Steerageway*, and while I do believe that you can impact every outcome through your efforts, it would be foolish for me to end before I say a word about luck. Luck is a lot like love: impossible to predict and difficult to describe, but most people know it when they see it; although, after the fact, almost all of us forget the luck and assume that it was our talent, insight, and unique goodness as a person that carried the day. Luck is hopelessly complicated, but luck happens to us all. I am very proud of my accomplishments, but I need to admit that I, like every successful leader, have had a lot of luck.

DEFINING LUCK

I have been lucky enough to see and touch foxfire, that mysterious, glowing material that is sometimes created when wood decays. Here is just how much luck had to go into that statement. I was lucky enough to be born at a time and in a place that allowed young people to be on their own a lot. I was lucky that I had a mother who was willing to support my adventures. For example, when I was fifteen years old she agreed to drop me off on the bank of the Kentucky River and to come back to get me two weeks later, 160 miles downstream. I was lucky enough to have a friend who was able to talk his mother into letting him go with me. I was lucky enough to find a canoe to use. I was lucky in picking the spot to be dropped off. I was lucky in not knowing how long it would take the two of us—brand new to canoeing—to make my goal of twenty miles in the first day, so we had to pick a campsite after dark. I was lucky that we picked a site that turned out to have an old log lying, half buried, right in the only place big enough to set up our tent. I was lucky to have brought an axe that I could use to chop off the top of that old rotten log. I was lucky that, unlike my friend who slept like a log, I rolled around in my sleep and actually rolled out of the tent (but not into the river). I was lucky to wake up in the middle of the night to find myself lying in the middle of all of those rotten wood chips, each one glowing with its own light. And, I was lucky I had a friend with me who I could wake up and who could see what I saw, because if he hadn't been there, I would have assumed that I was just going crazy and no one on earth would have believed this story.

Eleven amazing pieces of luck in support of that one unique experience. It makes it easy for me to appreciate all of the rest of the amazing luck that I have had along the way.

- I was lucky to have a mother who pushed me far beyond my comfort level. Without her, I would still be sitting on that riverbank.
- I was lucky to have a father who had an attitude. Without him, I would not have an emotional engagement in all that I do—and I wouldn't know any poetry.
- I was lucky to have a grandpa who started a little company and, above all else, loved to work.

FALLING IN LOVE WITH WORK

If you know me, you know my grandpa. I started going to work on construction projects with him and my uncles on the day after my sixth birthday; and no boy ever had such glorious adventures. There are two stories, among the many I can tell about Grandpa, that define my attitude toward work and that I believe are at the foundation of my success in life.

The first is about a long day when I was left all alone to dig fence-post holes. I cried and I cussed but I got the holes dug; and there was no feeling I had ever experienced like the feeling I had when Grandpa came back and said to me, "Good job, boy; I knew you could do it." I am happy with the job titles I have obtained, and I am satisfied with the financial results of my efforts; but every step of the way, I have worked in the hope that one more person would say, "Good job, boy; I knew you could do it."

The second story is a little more complicated. My fundamental job as a kid was to be a carpenter's helper. It was my job to get the tools out in the morning, to get the material to where the carpenters were working, to bring the nails, the screws, the next tool, and even an occasional dipper of water, so that those who knew better than I how to do the work were constantly engaged and effective. The job of helping others to be effective is the best preparation I can imagine for becoming a CEO. I learned early that there are people more talented than I at fitting a door but that I could contribute by making sure they had what they needed to do the job well. That lesson and the skills I developed in making sure that others have what they need to succeed are at the heart of my satisfaction and my success.

- I was lucky that Messer was the co-op job closest to Kentucky, when that was my only criteria for choosing a job.
- I was lucky that Messer had been getting steadily smaller for a decade so there was not a lot of competition to compare against my performance.
- I was lucky they didn't fire me when I complained about being assigned to the office.
- I was lucky I got sent to the Miami Valley Laboratories project, where I met Kathy.
- I was lucky that they didn't fire me for having long hair and a big mouth.
- I was lucky that I made such good friends at Messer—who cared about me when I wasn't smart enough to care about myself—so that I didn't accept the job that P&G offered.

HIRING ON

I told the story of Al Berndsen's hiring on with Messer. Now I will tell my own experience. Since I had heard nothing from Messer, even though I was working there part-time throughout my senior year at the University of Cincinnati, I signed up to interview with P&G. It was a wonderful experience—a full day of meeting delightful people and learning about a world that was completely new to me. At the end of the day, they offered me a job, and asked if I could get back to them as soon as possible. I almost said yes on the spot, since they sounded great, and I didn't have any other offers, but I did not, primarily out of fear that they would think that I was desperate.

The next day, I must have mentioned to Harry Ahlstrom, the project manager at the Miami Fort Station Project, that I was considering a job with P&G because the day after that Al Berndsen showed up at the project. Al leaned on the plan table next to me and asked what I intended to do after school. I gave some smart answer like, "I think I'll get a job." Al responded, "Why don't you work for us?" This drew the equally smart answer, "Because Messer hasn't offered me a job." He made the offer of two hundred dollars a week, which caused me to do some math and say, "Al, that's just 33 percent less than Procter & Gamble has offered me." Al then said the words that

won me over. He said, "Yes, but if you go to work for P&G you won't be working for me, will you?" I took the job for pretty much the same reason that he had taken the job almost twenty years earlier; I didn't want to let down the people I worked with.

- I was very lucky to have a wife who was both capable and willing to fill all of the gaps in my life, and who didn't care about that marginal 33 percent.
- I was lucky that Al Berndsen embraced growth, because that gave me the opportunity to be involved in hiring.
- I was lucky that Joe Glassmeyer decided to like me, because that gave me direction and influence.
- I was lucky that Larry Keckeis lost his battle to keep his sons out of construction.
- I was lucky that Bernie Suer came back for the fourth visit to try to get a co-op job after he was stood up on the first three visits.
- I was lucky that so many people across the company, after they heard my latest idea and rolled their eyes at my enthusiasm, were willing to make me a success through their efforts.
- And, I was lucky that so many gifted, caring people were willing to join Messer and to add their voices in building this new model.

And, while this list could go on and on, I feel like I shouldn't push my luck by asking you to read more.

Thank you all,
Pete Strange
www.messer.com
February, 2013

About the Author

MR. PETER S. STRANGE
CHAIRMAN
MESSER, INC., CINCINNATI, OHIO

Pete is Chairman of Messer, Inc., the parent company for Messer Financial Services and Messer Construction Co., a regional general contractor and construction manager. A graduate of the University of Cincinnati in Civil Engineering, he began his career with Messer as a co-op student. In 1989, he led negotiations resulting in the purchase of all outstanding Messer stock by management and employees.

Pete's test for leadership is simple: If you think you are a leader, look behind you. If there is no one there, you're just out for a stroll.

Pete and his wife, Ginger, reside in Erlanger, Kentucky. Their greatest accomplishments are their children—Andrew, Carrie, and Matthew and their grandchildren—Sam, Josh, Lily, Charley, and Harrison.

Acknowledgments

This book is not a creation; it is a recounting of special events and opportunities created by others. I am humbly grateful to all of those great builders who have spent their lives at Messer "turning dreams into reality at an epic scale." It is a blessing to be allowed to share in your accomplishments.

My special thanks to my assistant, Kim Spangler, who brought together my scraps of paper and shreds of ideas to provide useable pictures and print to our publishers.

My thanks to my several friends and associates who read the rough drafts and provided invaluable guidance.

Thanks to Richard Hunt, Donna Poehner, and Stephen Sullivan, who undertook the task of turning this literary sow's ear into something as close to a silk purse as possible.

And my special thanks to Tom, Kathy, and Bernie, who often did my work at Messer as well as their own, so I could be off trying something new like buying the company—or writing a book.

Finally, all of this—not just the book, but everything—is the result of one person, who has been willing and able to help me make better decisions, get better results, and think longer term. Thank you, Ginger; you are a true leader.

Capitalism at the individual level is the essence of the American dream.